# Think Before You Compute

Every fluid dynamicist will at some point need to use computation. Thinking about the physics, constraints and the requirements early on will be rewarded with benefits in time, effort, accuracy and expense. How these benefits can be realised is illustrated in this guide for would-be researchers and beginning graduate students to some of the standard methods and common pitfalls of computational fluid mechanics. Based on a lecture course that the author has developed over 20 years, the text is split into three parts. The quick introduction enables students to solve numerically a basic nonlinear problem by a simple method in just three hours. The follow-up part expands on all the key essentials, including discretisation (finite differences, finite elements and spectral methods), time-stepping and linear algebra. The final part is a selection of optional advanced topics, including hyperbolic equations, the representation of surfaces, the boundary integral method, the multigrid method, domain decomposition, the fast multipole method, particle methods and wavelets.

E. J. HINCH has been a teacher and researcher in fluid mechanics and applied mathematics at the University of Cambridge for over 45 years. He is the author of *Perturbation Methods* (Cambridge University Press, 1991) and has been awarded the Fluid Dynamics prizes of the European Mechanics Society and the American Physical Society Division of Fluid Dynamics.

# Cambridge Texts in Applied Mathematics

All titles listed below can be obtained from good booksellers or from Cambridge University Press. For a complete series listing, visit www.cambridge.org/mathematics.

# Think Before You Compute
## A Prelude to Computational Fluid Dynamics

E. J. HINCH
*University of Cambridge*

CAMBRIDGE
UNIVERSITY PRESS

# CAMBRIDGE
## UNIVERSITY PRESS

University Printing House, Cambridge CB2 8BS, United Kingdom

One Liberty Plaza, 20th Floor, New York, NY 10006, USA

477 Williamstown Road, Port Melbourne, VIC 3207, Australia

314–321, 3rd Floor, Plot 3, Splendor Forum, Jasola District Centre, New Delhi – 110025, India

79 Anson Road, #06–04/06, Singapore 079906

Cambridge University Press is part of the University of Cambridge.

It furthers the University's mission by disseminating knowledge in the pursuit of education, learning, and research at the highest international levels of excellence.

www.cambridge.org
Information on this title: www.cambridge.org/9781108479547
DOI: 10.1017/9781108855297

First published 2020

*A catalogue record for this publication is available from the British Library.*

ISBN 978-1-108-47954-7 Hardback
ISBN 978-1-108-78999-8 Paperback

# Contents

v

# Preface

This book, based on a graduate course in Cambridge, is aimed at students starting research into fluid mechanics who are thinking about computing a flow, as one amongst other tools of investigation. It is an educational book for beginners, using the simplest methods appropriate, rather than an advanced text for those already familiar with the methods. It is certainly not a research monograph about the very latest techniques. It is for those using a little computing for research in fluid mechanics. It is not for those researching into computational methods, either proving their mathematical properties or creating new methods.

The book is designed for students who have taken an undergraduate course on fluid mechanics and an undergraduate course on computing simple numerical methods, designed to lead those students to some understanding of computing flows. The course on fluid mechanics should have discussed the incompressible Navier–Stokes equation, the Reynolds number, boundary layers, vorticity and streamfunctions. The course on numerical methods should have included simple finite differencing of differential equations and iterative solutions. This book will then develop numerical methods appropriate to fluid mechanics. On the other hand, the book will not develop fluid mechanics. This means that no examples are included of numerical calculations in acoustics, aeronautics, compressible flows, combustion and reactions, biology, atmospheres, oceans, geology, non-Newtonian fluids and many industries. It is important to point out the many models of turbulence are also not included.

The book is divided into three parts. Part I is short and composed of three chapters. It tackles a very simple problem in fluid mechanics by very simple numerical methods. By making everything simple, students should be able to obtain results for a nonlinear flow after just one week of lectures. Some MATLAB code is available on my website,[1] so that students do not even have

---

[1] www.damtp.cam.ac.uk/user/hinch/teaching/CMIFM_Handouts/***.m, where *** is PoissonTest, StrfnVort and PrimVarb.

to spend time coding the programs themselves. But more than quick results, the first part delivers a far-from-hidden message of the need to think about what one is doing. There are issues of understanding the formulation of the question, of designing and monitoring the accuracy, of noting where time is consumed, of handling instabilities, of producing evidence that the answer is correct. There is also the special numerical issue in fluid mechanics of how to find the pressure. The simple problem tackled in Part I is the driven cavity, a square domain with a prescribed tangential velocity on the top surface. The simple numerical methods used are finite differences, central in space and forward in time, and successive over-relaxation of Gauss–Seidel iteration to solve a Poisson problem.

Part II gives a more detailed treatment of the general issues, such as turning a continuous partial differential equation into a finite discrete problem, i.e. discretisation by finite differences, finite elements and spectral methods, and general issues of time-stepping and solving large sparse systems of linear equations. Topics covered under discretisation include conservative formulations, a compact fourth-order Poisson solver, problems with pressure in finite elements, local vs global representations and the need for a pseudospectral approach. While time-stepping is only a discretisation in time, it deserves a more careful examination. There are issues of controlling the accuracy, not being too stable and sometimes avoiding excessive expensive evaluations of derivatives. While fluid mechanics is strictly nonlinear, large linear problems occur in the Poisson problem to find pressure or in considering the linear stability of a flow; hence the chapter on linear algebra. Students are strongly recommended not to code up finite elements or solvers for linear algebra but rather to use safe professionally written packages. The two chapters on these topics are included to explain what the packages are doing, so that the correct package can be used wisely.

Part III is an incomplete collection of specialised topics. The first chapter of Part III gives a quick introduction to one particular finite element package, FREEFEM++. I selected this package having tried several, because I have found that in less than an hour students can learn to use it to compute a flow. Solving hyperbolic equations numerically is unwise, and Chapter 10 illustrates the problems that arise with seemingly good schemes; only the one-dimensional case is presented. Some fluid mechanics problems involve moving boundaries. Chapter 11 discusses various representations of surfaces. This is followed by a chapter on the boundary integral method, which for potential flow and Stokes flow only uses data on the surface and so is highly suited for computing moving boundary problems. A Poisson problem typically consumes much of the time in computing a flow, so fast methods have been developed to solve it. For simple geometry, the multigrid method is probably the fastest, while in complex geometries domain-decomposition is particularly good with parallel computing. When the forcing of the Poisson problem is by many point sources, a fast multipole method can be useful, but only when there is a very large number of sources. While fluid mechanics is essentially about a continuum medium, particle descriptions naturally occur, whether one studies molecules moving in a gas, colloidal particles in a suspension, dry grains in a flowing granular medium or parcels of fluid in a Lagrangian description. Chapter 16 describes all these. The final chapter gives a quick introduction to wavelets, which have been found useful for analysing flows and identifying isolated regions of great activity.

A cautionary remark. One of the difficulties in computing flows is that every branch of fluid mechanics has its special physics, and that special physics should be reflected in some special numerics. Note the implication that there is no universal method or package applicable to every fluids problem. In fact I would go further to say that even for a particular problem there is no best method, one should always be able to dream up something better.

And finally I must acknowledge the enormous assistance of my colleagues Stephen Cowley, Paul Dellar and Paul Metcalfe in developing the graduate lecture course in Cambridge over a period of years.

# PART I

## A FIRST PROBLEM

# 1

# The driven cavity

## 1.1 The problem

We start with a simple problem for the Navier–Stokes equations, solved by simple methods. We will find the two-dimensional incompressible flow governed by

$$\nabla \cdot \mathbf{u} = 0,$$

$$\rho\left(\frac{\partial \mathbf{u}}{\partial t} + \mathbf{u} \cdot \nabla \mathbf{u}\right) = -\nabla p + \mu \nabla^2 \mathbf{u},$$

with constant density $\rho$ and viscosity $\mu$. These equations are to be solved inside a square $L \times L$ box with boundary conditions of no slip on the bottom and sides and a prescribed horizontal velocity along the top

$$\mathbf{u} = 0 \quad \text{on } y = 0 \text{ and } 0 < x < L, \text{ and on } x = 0 \text{ or } L \text{ and } 0 < y < L,$$

$$\text{and} \quad \mathbf{u} = (U(x), 0) \quad \text{on } y = L \text{ and } 0 < x < L.$$

This rectangular geometry is good for simple numerical methods.

We will evaluate the viscous force on the top

$$F = \int_0^L \mu \left.\frac{\partial u}{\partial y}\right|_{y=L} dx.$$

## 1.2 Know your physics

Before writing any code, it is worth thinking about the physics of the governing equations at the numerical grid level. The converse is also true that when presented with a new system of governing equations thinking about how to solve them numerically often deepens one's understanding of their physics.

3

The Navier–Stokes equations have three different physics activities represented by different combinations of the terms.

First on the left-hand side,

$$\frac{\partial \mathbf{u}}{\partial t} + \mathbf{u} \cdot \nabla \mathbf{u}$$

says that information is propagated with the flow, at velocity $\mathbf{u}$. This means in a short time interval $\Delta t$ information has propagated a distance $u\Delta t$ from one grid point towards another. One might want to limit the size of the time-step so that information is not propagated too far, say more than one space grid block, in one time-step.

Looking just at the far left and far right terms of the Navier–Stokes equations, we have

$$\rho \frac{\partial \mathbf{u}}{\partial t} = \mu \nabla^2 \mathbf{u},$$

i.e. a diffusion equation with a diffusivity of the kinematic viscosity $\nu = \mu/\rho$. Thus in one time-step $\Delta t$ information diffuses a distance $\sqrt{\nu \Delta t}$. Keeping this distance less than one grid block requires very small time-steps. While diffusion is relatively fast on small length scales, it is slow on a large length scales, so one often has to wait rather a long time for information to have diffused over the whole grid.

Finally the terms

$$\rho \frac{\partial \mathbf{u}}{\partial t} = -\nabla p \quad \text{with} \quad \nabla \cdot \mathbf{u} = 0$$

are capable of propagating information to great distances in zero time, particularly in impulsively started motion of bodies in a fluid, reflecting the infinite speed of sound in our incompressible fluid. This behaviour is awkward for numerical work, and an early warning that treating the pressure will not be easy.

The Reynolds number $Re = UL/\nu$ measures the relative importance of inertial to viscous terms in the Navier–Stokes equations. At low Reynolds numbers, typically $Re < 1$, vorticity diffuses rapidly and this must resolved numerically. On the other hand at high Reynolds numbers, typically $Re > 1,000$, there are thin boundary layers and sometimes long wakes which must be resolved numerically. To avoid these difficulties in our first simple problem, we shall set the Reynolds number to

$$Re = 10,$$

which is not too low and not too high. Moreover the analytical theories for low and for high Reynolds numbers will not work well at this intermediate value, making numerical solution the only way to solve the problem.

## 1.3 Know your PDEs

Before attempting to solve any equations, it is necessary to know what boundary and initial conditions must be satisfied in order to make the problem well posed, i.e. has a solution which is unique if appropriate and which is not too sensitive to the input data. Applying the wrong type of boundary conditions to an equation can result in there being no solution, although the computer will often misleadingly deliver an output.

The simplest partial differential equation, which is also present in the Navier–Stokes equations, is the first-order hyperbolic equation in one space and one time dimension

$$\frac{\partial \phi}{\partial t} + u(x,t)\frac{\partial \phi}{\partial x} = f(x,t).$$

To make this well posed one needs initial data $\phi(x,0)$ at $t = 0$ over some space interval $a < x < b$ along with inflow boundary data, say $\phi(a,t)$ at $x = a$ for $t > 0$ if $u(a,t) > 0$.

The next prototype equation is the second-order hyperbolic equation, better known as the wave equation

$$\frac{\partial^2 \phi}{\partial t^2} = c^2 \frac{\partial^2 \phi}{\partial x^2}.$$

Being second order in time, this needs both the initial value and the initial time derivative, $\phi(x,0)$ and $\phi_t(x,0)$ at $t = 0$ over some interval $a < x < b$. As information propagates in both directions, boundary data must be supplied at both ends of the interval, e.g. $\phi(a,t)$ and $\phi(b,t)$ for $t > 0$, although in place of the value the spatial derivative $\phi_x$ or some combination such as $\phi + \phi_x$ can be given. On an infinite domain, the boundary conditions are replaced by radiation conditions, which are often tricky to impose numerically.

Another second-order equation is an elliptic equation, better known as the Poisson or Laplace equation

$$\nabla^2 \phi = \rho.$$

This needs boundary data $\phi$ or $\partial \phi / \partial n$ or some combination $\alpha \partial \phi / \partial n + \beta \phi$ (with restrictions on $\alpha$ and $\beta$) given all around the boundary.

Finally there is the parabolic equation, better known as the heat equation

$$\frac{\partial \phi}{\partial t} = D\frac{\partial^2 \phi}{\partial x^2}.$$

Being first order in time, it needs initial data $\phi(x,0)$ at $t = 0$ over some interval, while being second order in space it needs information ($\phi$ or $\phi_x$ or a combination) at the boundaries at both ends $x = a$ and $x = b$.

The curious nomenclature comes from classifying the general linear second-order partial differential equation in two dimensions

$$a\frac{\partial^2\phi}{\partial x^2} + b\frac{\partial^2\phi}{\partial x\partial y} + c\frac{\partial^2\phi}{\partial y^2} + d\frac{\partial\phi}{\partial x} + e\frac{\partial\phi}{\partial y} + f = 0,$$

by comparing with the conic sections

$$ax^2 + bxy + cy^2 + dx + ey + f = 0$$

of hyperbolas, ellipses and parabolas. With the obvious exception of the degenerate case of a parabolic equation, the first- and zero-order derivatives play a minor role in determining the mathematical behaviour, and so can be nonlinear without changing what constitutes a well-posed problem.

Numerically, hyperbolic equations are the most difficult to solve. In that they preserve information which they propagate around, any numerical error will be preserved, until it accumulates to swamp the real solution. Elliptic equations are the most costly to solve numerically, because every point in the domain influences every other point, which produces a very large coupled problem. Parabolic equations are the easiest and cheapest equation to solve on a computer. Practically any method works and works well, because little numerical errors made at one time-step decay very rapidly within a few further steps.

## 1.4 Special physics of the corner

In the computational fluid dynamics (CFD) literature it is very common to take a constant uniform velocity along the top lid of the cavity

$$U(x) = U_0.$$

Unfortunately this has a stress singularity in the corners like $\sigma \propto r^{-1}$ due to the discontinuity of the velocity at the corners. The stress singularity gives an infinite force on the top plate.

A better choice would be a velocity which vanishes linearly into the corners

$$U(x) = U_0\sin(\pi x/L).$$

The viscous stresses are now regular, but the pressure has a logarithmic singularity. This weak singularity is integrable, but still is difficult to represent numerically.

Hence we shall take a velocity of the lid which vanishes quadratically at the corners

$$U(x) = U_0\sin^2(\pi x/L).$$

## 1.5 Nondimensionalisation

Engineers always use dimensional variables in computations: scientists do not.[1] We therefore scale the velocity $\mathbf{u}$ with $U_0$, lengths $x$ and $y$ with $L$, time $t$ with $L/U_0$ and pressure $p$ inertially with $\rho U_0^2$. This introduces a single non-dimensional group, the Reynolds number

$$Re = \frac{\text{inertial terms } \rho U_0^2/L}{\text{viscous terms } \mu U_0/L^2} = \frac{U_0 L}{\nu}.$$

The nondimensionalised problem is then

$$\nabla \cdot \mathbf{u} = 0,$$

$$\left( \frac{\partial \mathbf{u}}{\partial t} + \mathbf{u} \cdot \nabla \mathbf{u} \right) = -\nabla p + \frac{1}{Re} \nabla^2 \mathbf{u},$$

subject to boundary conditions

$$\mathbf{u} = 0 \quad \text{on } y = 0 \text{ and } 0 < x < 1, \text{ and on } x = 0 \text{ or } 1 \text{ and } 0 < y < 1$$

$$\text{and} \quad \mathbf{u} = (\sin^2(\pi x), 0) \quad \text{on } y = 1 \text{ and } 0 < x < 1.$$

We take a state of rest as the initial condition

$$\mathbf{u}(x, y, 0) = 0 \quad \text{at } t = 0 \text{ for } 0 < x < 1 \text{ and } 0 < y < 1.$$

We seek a solution at $Re = 10$.

Finally the force on the lid is scaled viscously by $\mu U_0$, so that we will evaluate

$$F = \int_0^1 \left. \frac{\partial u}{\partial y} \right|_{y=1} dx.$$

## 1.6 Steady vs transient calculations

While one might be interested only in the final steady state, it is normally easier to compute the time evolution from a simple initial condition to the final steady state. This is because the equations for the steady state are nearly always highly nonlinear, whereas the initial value problem is linear in the highest time derivative, e.g. linear in $\partial \mathbf{u}/\partial t$ in the Navier–Stokes equations. Moreover, there is the possibility that a steady state might not exist, or if it does exist might not

---

[1] An issue of philosophy. Engineers are interested in one practical realisation with all minor complications included, while scientists are interested in the general behaviour in a highly simplified model stripped of all minor complications. Adding complications increases the number of nondimensional groups faster than the number of dimensional variables.

be stable. The initial value problem will always have a solution, subject to there not being a finite-time blowup, and will show that a steady state is unstable if it is unstable.

Sometimes initial value problems approach the final steady state very slowly, and in such cases ways can be found to accelerate the slowly decaying transients.

Note that if one is interested in a series of steady problems, say the steady-state force on the top lid as a function of Reynolds number, then it is not necessary to begin each calculation from rest. Instead one can start the calculation for the next Reynolds number from the steady solution for the last Reynolds number. That would be a crude form of 'parameter continuation'.

Some relaxation methods for finding directly the steady state can be viewed as pseudotime evolutions.

§15.1 in Part III discusses methods for finding steady states.

## 1.7  Pressure!

The general idea for computing the evolution of the flow will be to be given $\mathbf{u}(x, t)$ at one time $t$, from this to evaluate $\partial \mathbf{u}/\partial t$ then and hence calculate $\mathbf{u}(x, t)$ at the next time level $t + \Delta t$. In this scheme we can easily evaluate the contributions to $\partial \mathbf{u}/\partial t$ from $-\mathbf{u} \cdot \nabla \mathbf{u}$ and from $\frac{1}{Re} \nabla^2 \mathbf{u}$. The problem arises of how we are going to find the pressure gradient $-\nabla p$. In analytic calculations, the pressure field just seems to drop out of the calculation, so that it is only when one first tries to find a flow numerically one realises that it is a nontrivial issue to find the pressure.

The pressure field enables one to satisfy the conservation of mass: mathematically speaking, it is the 'Lagrangian multiplier' associated with the solenoidal constraint $\nabla \cdot \mathbf{u} = 0$. In compressible fluids, the pressure is determined locally by the local density and temperature from an equation of state. In the incompressible limit, the pressure has to be determined globally by the need to make the velocity field solenoidal globally.

There are two alternative ways of tackling the pressure problem. In the so-called primitive variable formulation, we shall find the pressure gradient which makes the velocity solenoidal. Before tackling the problem head on, we will sidestep the pressure problem with the so-called streamfunction-vorticity formulation. This formulation is restricted to two-dimensional problems. The two formulations are taken up in the next two chapters.

# 2

# Streamfunction-vorticity formulation

## 2.1 Formulation

To satisfy automatically the solenoidal constraint on the two-dimensional velocity, $\nabla \cdot \mathbf{u} = 0$, we represent the velocity with a streamfunction $\psi(x, y, t)$

$$u = \frac{\partial \psi}{\partial y} \quad \text{and} \quad v = -\frac{\partial \psi}{\partial x}.$$

In this two-dimensional flow, the vorticity is

$$\omega = \frac{\partial v}{\partial x} - \frac{\partial u}{\partial y} = -\nabla^2 \psi.$$

We eliminate the troublesome pressure by taking the curl of the momentum equation, so forming the vorticity equation

$$\frac{\partial \omega}{\partial t} + \mathbf{u} \cdot \nabla \omega = 0 + \frac{1}{Re} \nabla^2 \omega.$$

Note that there is no vortex-stretching term for the two-dimensional flow. The advection of vorticity can be written alternatively

$$\mathbf{u} \cdot \nabla \omega = \psi_y \omega_x - \psi_x \omega_y = \frac{\partial(\omega, \psi)}{\partial(x, y)}.$$

Two boundary conditions must be applied. The condition for no normal component of velocity on all the sides makes the sides a streamline, and hence we can impose

$$\psi = 0.$$

9

The boundary condition on the tangential velocity is

$$\frac{\partial \psi}{\partial y} = \sin^2 \pi \quad \text{on the top, } y = 1 \text{ and } 0 < x < 1,$$

$$\frac{\partial \psi}{\partial y} = 0 \quad \text{on the bottom, } y = 0 \text{ and } 0 < x < 1,$$

$$\frac{\partial \psi}{\partial x} = 0 \quad \text{on the sides, } x = 0 \text{ or } 1 \text{ and } 0 < y < 1.$$

We will solve the problem treating it as a decoupled pair of problems. First we shall solve a Poisson problem, of given the vorticity $\omega$ at time $t$ solve for the streamfunction $\psi$ at that time,

$$\nabla^2 \psi = -\omega, \quad \text{subject to } \psi = 0 \text{ on the boundary.}$$

Then knowing the streamfunction and the vorticity at time $t$, we can evaluate the time derivative of the vorticity

$$\frac{\partial \omega}{\partial t} = -\frac{\partial(\omega, \psi)}{\partial(x, y)} + \frac{1}{Re} \nabla^2 \omega,$$

and so calculate the vorticity at the next time-step. The slight complication in this attractive idea of decoupling the problem is that the second boundary condition is on the normal derivative of the streamfunction and not the vorticity: we will have to choose the vorticity on the boundary so that the normal derivative of the streamfunction takes the correct imposed value.

## 2.2  Finite differences (simple)

On a finite computer we must work with a finite representation of the (infinite) continuous functions $\psi(x, y, t)$ and $\omega(x, y, t)$. In the simplest treatment of the simple driven-cavity problem, we will use a simple finite difference representation. In Part II, 'Generalities', we give more sophisticated finite difference representations and alternative representations by finite elements and spectral functions.

For finite differences, we hold values of the unknown functions at a discrete number of times at a discrete number of points on a grid or mesh. The simplest is at equally spaced time intervals $\Delta t$ over equally spaced $\Delta x$ points on a square, $\Delta y = \Delta x$, i.e.

$$\omega_{ij}^n \approx \omega(x = i\Delta x, y = j\Delta x, t = n\Delta t).$$

The approximation sign is a reminder that while we would like to know the

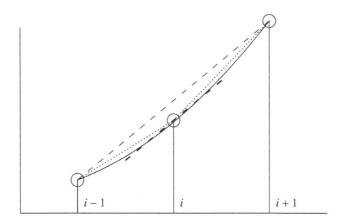

Figure 2.1 Estimates of the derivative at $i$ (dashed line through $i$), by forward differencing to $i + 1$ (dotted line through $i + 1$), backward differencing to $i - 1$ (dotted line through $i - 1$) and central differencing between $i - 1$ and $i + 1$ (dashed line through $i - 1$ and $i + 1$). The positive curvature makes the first too large and the second too low.

unknown function exactly we will in fact find only a numerical approximation to it at these points and times.

As we are solving partial differential equations, we need approximations to derivatives from the above representations. For a function of one variable, $f(x)$, we can find an approximation to the derivative $f'$ at $x = i\Delta x$, i.e. $f_i'$, from the change in value from nearby points. There are several possibilities, see Figure 2.1:

$$\text{Forward differencing } f_i' = \frac{f_{i+1} - f_i}{\Delta x} + O(\Delta x),$$

$$\text{Backward differencing } f_i' = \frac{f_i - f_{i-1}}{\Delta x} - O(\Delta x),$$

$$\text{Central differencing } f_i' = \frac{f_{i+1} - f_{i-1}}{2\Delta x} + O(\Delta x^2).$$

The approximations would be exact if the function were linear. A quadratic part of the function leads to a first-order error $O(\Delta x)$ of opposite sign in the forward and backward differences, and a second-order error $O(\Delta x^2)$ in the central difference. If possible, one uses the more accurate central differencing.

The second derivative $f''$ of the function can be estimated similarly from the above approximations for the first derivatives at fictitious mid- or half-points.

$$f_i'' \approx \frac{\left(f_{i+\frac{1}{2}}' \approx \frac{f_{i+1} - f_i}{\Delta x}\right) - \left(f_{i-\frac{1}{2}}' \approx \frac{f_i - f_{i-1}}{\Delta x}\right)}{\Delta x}$$

$$= \frac{f_{i+1} - 2f_i + f_{i-1}}{\Delta x^2} + O(\Delta x^2).$$

Note that the above approximation to the second derivative is not the same as the approximate first derivative of the approximate first derivative (all by central differencing)

$$f_i'' \neq (f_i')' = \frac{f_{i+2} - 2f_i + f_{i-2}}{4\Delta x^2}.$$

While both expressions for the second derivative have second-order errors, the errors tend to be four times larger in the last, less compact form. There is a similar problem with the product rule for approximations to the derivative,

$$(ab)' \neq a'b + ab'.$$

Sometimes, e.g. when constructing 'conservative schemes', it can be important to use the left-hand form rather than the right.

One can find the leading order error in the above numerical approximations to derivatives using Taylor series expansions

$$f_{i+1} = f(x = i\Delta x + \Delta x)$$
$$= f_i + \Delta x f_i' + \frac{1}{2}\Delta x^2 f_i'' + \frac{1}{6}\Delta x^3 f_i''' + \frac{1}{24}\Delta x^4 f_i'''' + \cdots.$$

Hence for the approximation to the second derivative

$$f_{i+1} - 2f_i + f_{i-1} = \Delta x^2 f_i'' + \frac{1}{12}\Delta x^4 f_i'''' + \cdots.$$

This justifies the error estimate $O(\Delta x^2)$ in the first expression for second derivative above, so long as $f''''$ is finite.

We will be using central differencing, so producing $O(\Delta x^2)$ errors, in our spatial derivatives. We shall find that the less accurate forward differencing will be adequate for time derivatives in our simple driven-cavity problem.

Adding the second derivative in $x$ to the second derivative in $y$, we form a numerical approximation to the Laplacian,

$$\left(\nabla^2 \psi\right)_{ij} \approx \frac{\psi_{i+1\,j} - 2\psi_{ij} + \psi_{i-1\,j}}{\Delta x^2} + \frac{\psi_{i\,j+1} - 2\psi_{ij} + \psi_{i\,j-1}}{\Delta x^2},$$

which can be nicely written with a 'numerical molecule'

$$\approx \frac{1}{\Delta x^2} \begin{pmatrix} & 1 & \\ 1 & -4 & 1 \\ & 1 & \end{pmatrix} \psi_{ij}.$$

## 2.3 Poisson problem

The first part of the decoupled streamfunction-vorticity formulation is to solve for the streamfunction $\psi$ given the vorticity $\omega$,

$$\nabla^2 \psi = -\omega.$$

We are solving this equation in a $1 \times 1$ box. If we have $N$ equally spaced points in the $x$ and $y$ directions, then the space-step $\Delta x = 1/N$.

At interior points, $i = 1 \to N - 1, j = 1 \to N - 1$, we solve

$$\frac{1}{\Delta x^2} \begin{pmatrix} & 1 & \\ 1 & -4 & 1 \\ & 1 & \end{pmatrix} \psi_{ij} = -\omega_{ij},$$

with boundary conditions

$$\psi = 0 \quad \text{for} \quad i = 0 \; \& \; N, \quad j = 0 \to N \quad \text{and for} \quad j = 0 \; \& \; N, \quad i = 0 \to N.$$

This is just a linear algebra problem for the unknown $\psi_{ij}$, but quite a large linear algebra problem. On a small $10 \times 10$ grid, there are 64 equations for 64 unknowns, i.e. the problem could be viewed as inverting a $64 \times 64$ matrix. However, most of the entries in that matrix are zeros, i.e. the matrix is sparse. Direct inversion of the matrix is expensive, particularly for larger grids, so we are interested in faster iterative methods of obtaining a good approximation to the solution for $\psi_{ij}$. Even with the faster iterative method described below, over 90% of the CPU will be spent on solving the Poisson problem. This is typical of computing fluid flows. Hence in Part II, 'Generalities', we need to return to faster ways of solving sparse problems in linear algebra, and in Part III, Chapter 13 describes several fast Poisson solvers.

The simplest iterative method for solving for the unknown $\psi_{ij}$ is the Gauss–Seidel method. One sweeps through the interior points in a logical order, first with $j = 1$ and $i = 1 \to N - 1$, then with $j = 2$ and $i = 1 \to N - 1 \ldots$ up to $j = N - 1$ and $i \to N - 1$, see Figure 2.2. This is a single sweep through the grid which has to be repeated many times. At each point of the sweep one replaces

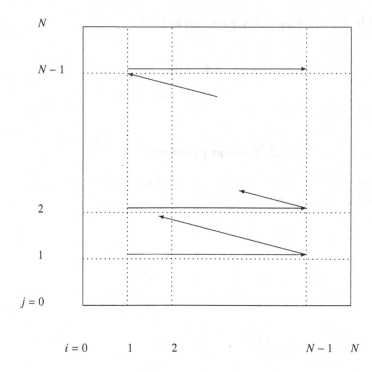

Figure 2.2  Gauss–Seidel sweeping through the grid.

$\psi_{ij}$ with

$$\psi_{ij}^{\text{new}} = \frac{1}{4} \left( \psi_{i+1\,j}^{\text{old}} + \psi_{i-1\,j}^{\text{new}} + \psi_{i\,j+1}^{\text{old}} + \psi_{i\,j-1}^{\text{new}} + \Delta x^2 \omega_{ij} \right).$$

Note in Figure 2.3 that the value on the row below $\psi_{i,j-1}$ and the value to the left $\psi_{i-1\,j}$ have just been modified in the current sweep, while the value on the row above $\psi_{i,j+1}$ and the value to the right $\psi_{i+1\,j}$ will be modified later, except when any of these values are boundary values. One needs about $N^2$ sweeps in order for the iterative process to converge. As each sweep involves $N^2$ unknown values $\psi_{ij}$, this iterative solution of the Poisson problem takes $O(N^4)$ total operations. This should be contrasted with $O(N^6)$ operations for the direct inversion of the $N^2 \times N^2$ matrix.

A little better than the Gauss–Seidel iteration, but not 'the' best by a long way, is successive over-relaxion (SOR). This modification of Gauss–Seidel is very simple to program so well worth the extra line of code. With a relaxation

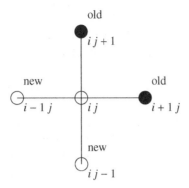

Figure 2.3 Gauss–Seidel use of new values from current sweep and old values from previous sweep.

parameter $r$, SOR iterates

$$\psi_{ij}^{\text{new}} = (1 - r)\psi_{ij}^{\text{old}} + r\{\text{above expression for } \psi_{ij}^{\text{new}}\}.$$

The iteration process is called 'under-relaxation' if the parameter is in the range $0 < r < 1$, it is the original Gauss–Seidel if $r = 1$ and is 'over-relaxation' if $1 < r < 2$. The iteration diverges if $r \geq 2$. There is an optimal value of the relaxation parameter which depends on the problem and on the size of the grid. For our Poisson problem and for large $N$, the optimal value is

$$r_{\text{optimal}} = \frac{2}{1 + \pi/N}.$$

With this optimal value one needs only $O(N)$ sweeps, rather than $N^2$ sweeps, in particular $2N$ sweeps give four-figure accuracy for $\psi$.

A disadvantage of the SOR method is the need to know the optimal value of the relaxation parameter. While there may not be a theoretical prediction for other problems, simple numerical experimentation often quickly finds the optimal value which can then be used many times, say every time-step.

## 2.4 Test the code

All computer codes have bugs when first written, probably several bugs. It is therefore advisable to write code in small subroutines, and first test thoroughly each subroutine in isolation.

A common error is for the indices to sweep wrongly through the grid. A first check then is to turn on the 'range-checking' option of the compiler in order to see if the code is trying to access a variable outside the range it should. One can also try compiling the code with two different compilers, because different compilers often have different bad behaviours when wrongly addressing arrays. Note some compilers object if one evaluates an expression using values of a variable which have not previously been set. It is good practice to initialise variables before using them, whether or not a particular compiler requires it, because one then obtains reproducible results.

Next it is well worth checking the null case. Check if one sets $\omega = 0$ everywhere, then the code delivers the correct answer $\psi = 0$ everywhere.

Now we need a test case where we know the exact theoretical solution. For our Poisson problem, a good test case is to set

$$\omega(x, y) = 2\pi^2 \sin \pi x \sin \pi y,$$

which should produce a good approximation to the known solution

$$\psi(x, y) = \sin \pi x \sin \pi y.$$

First one plots a graph of the numerical solution $\psi(x, y)$, see Figure 2.4. Is the shape correct? Figure 2.4 shows a function vanishing on the boundaries with a single maximum in the centre of the interior. Is the solution nonzero throughout the interior, or is the iterative sweep missing a line? Are the magnitudes correctly $O(1)$ or is the code missing a factor of $\Delta x^2$? Figure 2.4 shows the central maximum is about 1.

Next we need to study the rates of convergence of the iterative scheme. Figure 2.5 shows how the value of the streamfunction in the middle of the grid $\psi(0.5, 0.5)$ converges with the number of iterations for various values of the relaxation parameter $r$. When $r = 1$ the relaxation is Gauss–Seidel, which takes about $N^2 = 400$ iterations to converge to four-figure accuracy. The convergence is much faster for the optimal value $r = 1.75$, needing less than $\frac{3}{2}N$ for the same convergence. A little higher than the optimal value, $r = 1.8$, overshoots the result, while a little lower approaches monotonically but slightly slower.

The final test is to see if the code which was designed to satisfy the equation with local errors $O(\Delta x^2)$ does really agree with the known solution with errors

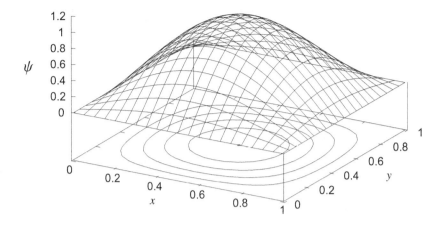

Figure 2.4 The solution for the test streamfunction for $N = 20$.

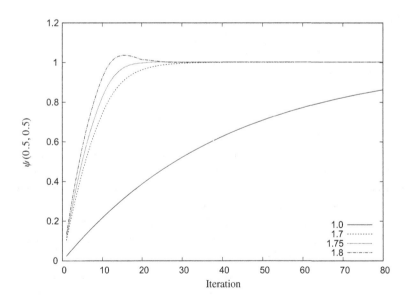

Figure 2.5 Convergence of the SOR iterations for $N = 20$ and for different values of the relaxation parameter $r$.

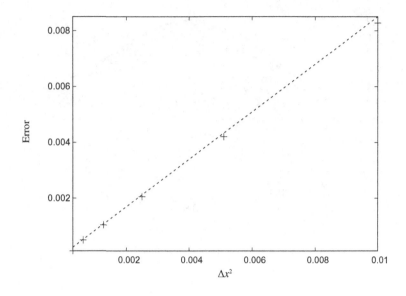

Figure 2.6  Test for second-order error, from $N = 14, 20, 28, 40$ and $56$.

of that size. Figure 2.6 gives the maximum error over the grid,

$$\text{Error} = \max_{\text{grid}} \left| \psi_{ij}^{\text{numerical}} - \psi^{\text{theory}}(i\Delta x, j\Delta) \right|,$$

plotted against $\Delta x^2$. There is a linear relation between the error and $\Delta x^2$, confirming the correct second-order behaviour. The line corresponds to $0.85\Delta x^2$. Thus for this test problem, $N = 10$ will give a 1% accurate answer, which is an acceptable accuracy when working quickly. If there is need for high three-figure accuracy, $N = 28$ is sufficient. The CPU cost of the refined $N = 28$ grid is over 20 times that of the course $N = 10$ grid. Having established a quadratic decay in the error, one can use this behaviour to extrapolate to $N = \infty$ from a couple of cheap coarse grids.

Our test of the Poisson solver was chosen so that we had an exact theoretical solution against which to test the numerical solution. For nonlinear problems, such as the full driven-cavity problem at $Re = 10$, one does not have the luxury of a known theoretical test solution. In such cases one can still test that a typical value of the solution, such as the midpoint value $\psi(0.5, 0.5)$ which we used earlier, converges to the unknown answer quadratically in $\Delta x$.

## 2.5 Code quality

There are two extreme styles of writing code, one-off code and production code. Normally one foolishly lives between these two extremes combining the worse features of both.

One-off codes will be written one day, used that day and never again. The programmes need to be small, very simple, with a very clear layout, and to contain no fancy tricks.

Production codes will be used many times, possibly by others, often by one-self in several years time after a break when one forgets how it all works. They need the following.

1. Comments to most lines. One needs to give the purpose and action of each morsel of code. One can also indicated possible future developments.
2. To fail gracefully. One needs to test for possible problems, e.g. failure to converge (no infinite loops), printing out a helpful message and giving useful values of variables before halting. Certainly the program must not continue to run ignoring the occurrence of a problem in some hidden sub-routine.
3. Subroutines that are bulletproof. Implementing this depends on the programming language, but one would like to ensure that a subroutine alters only those external variables that it is designed to alter, with no indirect or accidental modifications of other external variables.
4. To be fast and efficient. There are little tricks such as avoiding repeating exactly the same calculation every iteration. Thus by setting $r1 = 1 - r$, $r025 = 0.25r$ and $h2w_{ij} = h^2\omega_{ij}$, one can write the SOR relaxation as

$$\psi_{ij} = r1\psi_{ij} + r025\left[\begin{pmatrix} & 1 & \\ 1 & & 1 \\ & 1 & \end{pmatrix}\psi_{ij} + h2w_{ij}\right],$$

which involves only two multiplications in place of five before. Note the loss of clarity which accompanies these tricks. Modern optimising compilers often automatically make such rearrangements.

Professional packages are finely tuned to a particular machine. If there is such a package, do use it. For example, there are often routines to multiply two matrices or to form the scalar product between two vectors. These can be 10 times faster than hand coding the same calculation, because the special package can fetch data optimally from slower memory to produce a pipeline which yields one result every cycle of the chip. At a higher level, there are always good, safe, fast packages for linear algebra.

## 2.6 Simple graphs

A very important tool when computing flows is the plotting of graphs. A visual picture is so much easier to interpret than tables of numbers. One always has to worry that the code is not behaving as designed. So, is the picture correct? The simplest way of quickly producing a graph is the following using the public domain GNUPLOT. There are many better, but more complicated ways.

First, one arranges for the programme to write out (to the default standard output stream) a table of results, i.e. on the first line $x_1$ $y_1$, on the next $x_2$ $y_2$, etc. That is for producing a line plot $y(x)$. For a contour plot of $z(x, y)$, one has three entries on each line $x_1$ $y_1$ $z_1$, etc.

Next, one streams this output to screen into a file called, say, *res* with the command line (under unix): *a.out > res*. Then one calls the software *gnuplot*, which responds with a > prompt.

For line plots $y(x)$ one types after the prompt:

> *plot 'res' with lines*

The *with lines* can abbreviated to *w l*. The scales for the graph are automatically adjusted, but can be set manually. One can plot several functions on the same graph, use log scales, set labels for the graph and axes and more.

For contour plots $z(x, y)$, one types after the prompt

> *splot 'res'*

Sometimes it is useful to issue first the commands *set parameter* and then *set contour*. Again there are many options.

To find out about the many extras, one can type after the prompt

> *help plot*

Finally to finish, one types *quit* after the prompt.

## Exercises

Part I of this book has been designed to enable new students to compute a simple nonlinear flow with one week of lectures. So it is now time to start computing. MATLAB code is available on my website to solve the simple Poisson problem, but it would be better to write your own code.

**Exercise 2.6.1**  Using the approach described in §2.3, solve the Poisson equation

$$\nabla^2 \psi = -2\pi^2 \sin \pi x \sin \pi y \quad \text{in } 0 \le x \le 1, 0 \le y \le 1,$$

with boundary condition $\psi = 0$ on $x = 0$ and $1$ and $y = 0$ and $1$.

**Exercise 2.6.2** Examine the graph of the solution for $\psi$ to see if it is sensible and like Figure 2.4.

**Exercise 2.6.3** Find how the solution for the middle of the grid $\psi(0.5, 0.5)$ converges with the number of iterations. Change the value of the SOR parameter $r$. Is the behaviour like that in Figure 2.5?

**Exercise 2.6.4** Find how everything depends on the number of points $N$.

**Exercise 2.6.5** The final test is to compare the numerical answer at different resolutions $N$ with the exact analytic solution

$$\psi = \sin \pi x \sin \pi y.$$

Are the local errors $O(\Delta x^2)$, as in Figure 2.6?

## 2.7 Vorticity evolution

We now turn to the equation for how the vorticity evolves in time

$$\frac{\partial \omega}{\partial t} = -\frac{\partial(\omega, \psi)}{\partial(x, y)} + \frac{1}{Re}\nabla^2\omega.$$

We start from an initial condition of rest

$$\omega(x, y, t) = 0 \quad \text{at } t = 0.$$

Using central differencing in space and simple forward differencing in time, we step from time level $n$ to time level $n+1$ at interior points $i = 1 \rightarrow N-1, j = 1 \rightarrow N - 1$,

$$\omega_{ij}^{n+1} = \omega_{ij}^n$$
$$- \Delta t \left( \frac{\omega_{i+1\,j}^n - \omega_{i-1\,j}^n}{2\Delta x} \frac{\psi_{i\,j+1}^n - \psi_{i\,j-1}^n}{2\Delta x} - \frac{\omega_{i\,j+1}^n - \omega_{i\,j-1}^n}{2\Delta x} \frac{\psi_{i+1\,j}^n - \psi_{i-1\,j}^n}{2\Delta x} \right)$$
$$+ \frac{\Delta t}{Re\Delta x^2} \begin{pmatrix} & 1 & \\ 1 & -4 & 1 \\ & 1 & \end{pmatrix} \omega_{ij}^n.$$

Note that at the interior points next to the boundary this expression calls for values of $\psi$ and $\omega$ on the boundary. Now on the boundary we know $\psi = 0$. However we do not know $\omega$ on the boundary but rather the normal derivative of the streamfunction is equal to the imposed tangential velocity

$$\frac{\partial \psi}{\partial n} = U_{\text{wall}}.$$

Consider the bottom boundary $y = 0$. By central differencing the streamfunction we can find a numerical approximation to the tangential velocity one half-step above the boundary

$$u_{\frac{1}{2}} = \frac{\psi_{i1} - \psi_{i0}}{\Delta x}.$$

By central differencing of this velocity at the half-point with the given wall velocity, we can find a numerical approximation to the vorticity at the quarter-point above the boundary

$$\omega_{\frac{1}{4}} = \frac{u_{\frac{1}{2}} - U_{\text{wall}}}{\frac{1}{2}\Delta x}.$$

The crudest approximation for a boundary condition on the vorticity would therefore be to set the vorticity on the boundary equal to its value at the quarter-point above

$$\omega_0 \approx \omega_{\frac{1}{4}} = \frac{\dfrac{\psi_{i1} - \psi_{i0}}{\Delta x} - U_{\text{wall}}}{\frac{1}{2}\Delta x}.$$

This approximation has first-order $O(\Delta x)$ errors. A better boundary condition on the vorticity is to use linear extrapolation from the quarter-point and one full point in the interior to obtain the second-order condition

$$\omega_0 = \frac{4\omega_{\frac{1}{4}} - \omega_1}{3}.$$

Starting from rest $\psi = 0$, the vorticity is initially very high on the boundary, $\omega_0 = -U_{\text{wall}}/\frac{1}{2}\Delta x$. This is effectively a numerical delta function. In time the vorticity diffuses over the interior until the boundary condition $\partial \psi / \partial n = U_{\text{wall}}$ is satisfied smoothly.

## 2.8 Time-step instability

If $\Delta t$ is too large, there is a numerical instability, clearly seen in Figure 2.7 of $\omega$ at $t = 0.525$ with $\Delta t = 0.035$, $\Delta x = 0.1$ and $Re = 10$.

When one first sees a graph like Figure 2.7, one has to worry whether the instability is numerical or physical. Now we are supposed to understand the physics of the problem before tackling it numerically. From that understanding, we expect no physical instability at the low value of the Reynolds number $Re = 10$. Perhaps by $Re = 200$ the flow becomes unstable, but not for low values in this geometry. Hence we guess that the instability has a numerical

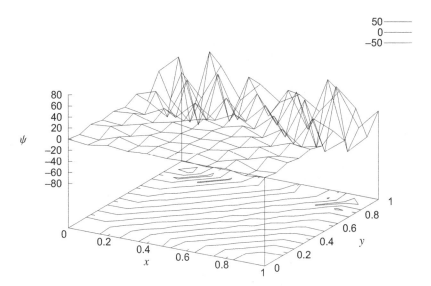

Figure 2.7 Numerical instability in time-stepping; a checkerboard form. The vorticity $\omega$ for $Re = 10$ at $t = 0.525$ with $\Delta t = 0.035$ and $\Delta x = 0.1$.

cause. Further, the checkerboard form of opposite signs at adjacent points is very typical of numerical instabilities.

The numerical instability comes from the diffusion terms. If one assumes that the instability takes the checkerboard form of opposite signs at adjacent points

$$\omega_{ij}^n = (-)^{i+j} A_n,$$

|  | + | − | + |  |
|---|---|---|---|---|
|  | − | + | − |  |
|  | + | − | + |  |

and substitutes this into the diffusional part of the time-stepping algorithm, one finds

$$A_{n+1} = A_n + \frac{\Delta t}{Re\Delta x^2} \cdot - 8A_n$$

and so

$$A_n = \left(1 - \frac{8\Delta t}{Re\Delta x^2}\right)^n A_0.$$

This blows up if the bracket has a modulus greater than unity. Thus for stability, we need

$$\Delta t < \tfrac{1}{4}Re\Delta x^2.$$

In physical terms, one has to have the time-step smaller than the time to diffuse across one grid block. I tend to work with $\Delta t = \tfrac{1}{5}Re\Delta x^2$, i.e. at 80% of the stability boundary, which gives a large $\Delta t$ which is safely stable.

The advection terms can also give rise to a numerical instability. To avoid this one must satisfy the Courant–Friedricks–Lewy (CFL) condition

$$\Delta t < \Delta x/u_{max}.$$

In physical terms, one has to have the time-step smaller than the time to advect information across a grid block.

For our driven-cavity problem, we need a spatial resolution fine enough to resolve boundary layers. This requires the 'grid-Reynolds-number' to be small, i.e. in dimensional terms $u_{max}\Delta x/\nu < 1$ or in dimensionless terms

$$\Delta x < 1/Re.$$

If we choose the spatial resolution $\Delta x$ to satisfy this and choose the time-step $\Delta t$ to satisfy the diffusion stability condition, then the advection CFL condition is automatically satisfied.

The stability restriction on the size of $\Delta t$ is quite severe, leading to a rapid escalation in the CPU costs when spatial resolution is increased. The number of time-steps needed to reach the finite time $t = 1$ is $1/\Delta t \propto N^2$. Now each time-step costs $N^3$ operations from the SOR Poisson solver. Hence the total cost will be $O(N^5)$. Thus doubling $N$ takes 32 times longer to run.

There are more sophisticated time-stepping algorithms, which allow larger $\Delta t$, but which are less accurate with the larger $\Delta t$. In Part II, 'Generalities', we need to return to time-stepping algorithms.

## 2.9 Accuracy consistency

We now need to prove that the time-stepping component of the code is correct. For this transient nonlinear problem of the evolution of the vorticity we no longer have an analytic solution to test against. One check, although not really

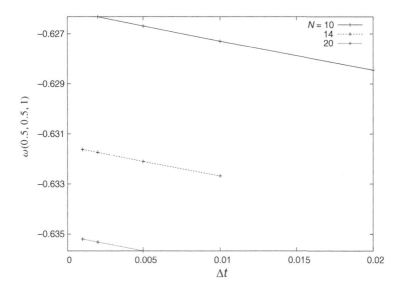

Figure 2.8 Test for $O(\Delta t)$ accuracy in $\omega(0.5, 0.5, 1)$ when applying the first-order boundary condition on the vorticity $\omega_0 \approx \omega_{\frac{1}{4}}$, with $Re = 10$ and $N = 10, 14$ and 20.

sufficient, is to demonstrate the numerical accuracy is that of the design of the algorithm, i.e. the errors are $O(\Delta t, \Delta x^2)$.

We will look at the vorticity $\omega$ at the midpoint $x = 0.5, y = 0.5$ at the finite time $t = 1$. Care is needed to ensure that one is exactly at $x = 0.5, y = 0.5, t = 1$ and not out by one $\Delta x$ or $\Delta t$; in particular, the number of grid points $N$ has to be even and $1/\Delta t$ has to be an integer.

Figure 2.8 gives the point vorticity $\omega(0.5, 0.5, 1)$, calculated applying the first-order boundary condition on the vorticity $\omega_0 \approx \omega_{\frac{1}{4}}$, as a function of the time-step $\Delta t$ for various spatial grids $N$. Note $\Delta t$ is limited by the stability condition, and so the maximum possible value decreases with $N$. The curves show a clear linear dependence in $\Delta t$, in fact the changes for each $N$ are given by $0.12\Delta t$. The errors from time-stepping are however rather small, about 0.001. There are much larger errors from the poorer spatial resolution, as can be seen by the gap between the three lines in Figure 2.8.

Now the time-stepping algorithm was designed with central spatial differencing to be second-order accurate in $\Delta x$, which the results in Figure 2.8 are not. A large first-order error was introduced by the first-order boundary condition $\omega_0 \approx \omega_{\frac{1}{4}}$. Improving this to the second-order boundary condition $\omega_0 \approx (4\omega_{\frac{1}{4}} - \omega_1)/3$, gives the results plotted in Figure 2.9. Note the vertical scale

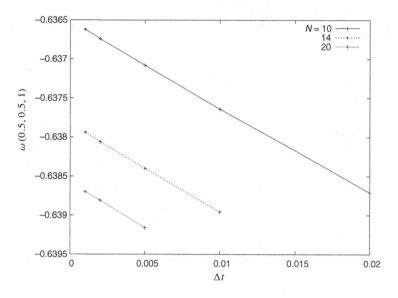

Figure 2.9 Test for $O(\Delta t)$ accuracy in $\omega(0.5, 0.5, 1)$ when applying the second-order boundary condition on the vorticity $\omega_0 \approx (4\omega_{\frac{1}{4}} - \omega_1)/3$, with $Re = 10$ and $N = 10$, 14 and 20.

in this figure is 30% of that in the previous figure, i.e. the spatial errors are much reduced. The spatial errors are now of a similar size to the temporal errors.

For diffusion-like problems such as our driven cavity, a *well-matched design* is a code which is second order in space and first order in time, with the time-step set at 80% of the stability condition, $\Delta t = 0.2Re\Delta x^2$. With that size time-step, the first-order time errors $O(\Delta t)$ are the same size as the second-order spatial errors $O(\Delta x^2)$. Hence for this problem, it is not worth building a more sophisticated second-order time-stepping scheme, nor taking $\Delta t$ much smaller than the stability boundary.

Having checked the temporal errors are $O(\Delta t)$, we finally have to check that the spatial errors are $O(\Delta x^2)$. We set $\Delta t = 0.2Re\Delta x^2$. Figure 2.10 plots the point vorticity as a function of $\Delta x^2$. We see a roughly linear decrease in $\Delta x^2$ up to $N = 24$. We can confidently claim that $\omega(0.5, 0.5, 1) = -0.63925 \pm -0.00005$. The curvature in Figure 2.10 at small $\Delta x$ may be due to difficulties in resolving the $t^{-1/2}$ concentration of vorticity at early times.

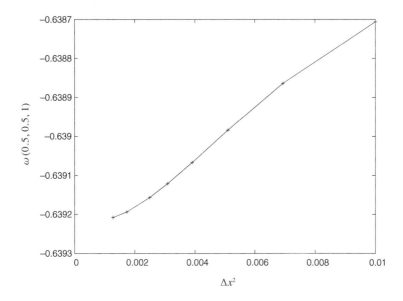

Figure 2.10 Test for $O(\Delta x^2)$ accuracy in $\omega(0.5, 0.5, 1)$ when applying the second-order boundary condition on the vorticity, with $Re = 10$ and $N = 10, 12, 14, 16,$ $18, 20, 24$ and $28$, and with $\Delta t = 0.2Re\Delta x^2$.

## 2.10 Results

After so much time developing and testing the code, it is time for a few results.

**Time to evolve.** Figure 2.11 shows the approach to the steady state of the vorticity in the middle of the box, $\omega(0.5, 0.5, t)$. We see by $t = 2$ the vorticity is steady to four significant figures. This is roughly the time for vorticity to diffuse across the box. The following results for the steady state are evaluated at $t = 3$.

If one is interested only in the steady state and not the time evolution, then one can try to reduce the computing time from $O(N^5)$ to $O(N^4)$ by only making a few, say three, iterations of the SOR Poisson solver each time-step, rather than the $2N$ iterations needed to converge fully. The few iterations need to be started from the result at the end of the last time-step, rather than starting from zero. As $15N^2$ time-steps are taken to get to $t = 3$, there will be more than the $2N$ iterations needed for the Poisson solver to converge. Note that this time-saving tactic does not always work smoothly.

**Steady streamfunction.** Figure 2.12 gives the steady-state streamfunction. One sees that the streamlines are crowded near the moving top at $y = 1$, indicating the highest velocities there. The flow is slow in the bottom half of the

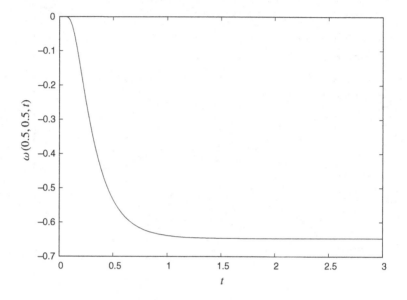

Figure 2.11 Evolution to the steady state for the vorticity in the centre of the box with $N = 20$ and $Re = 10$.

box and near the no-slip sidewalls. At $Re = 10$ the streamfunction retains the symmetry around the centreline $x = \frac{1}{2}$ of the low Reynolds number limit. The bottom corners show a small region, less than 0.1 in size, of reversed corner eddies. The reversed circulation is very weak, about $2 \ 10^{-6}$ compared to the main circulation of $8 \ 10^{-2}$.

**Steady vorticity.** The streamfunction is necessarily very smooth, and as a consequence shows little of the structure of the solution. Much better is the steady-state vorticity in Figure 2.13. The vorticity is largest next to the moving lid, and is negative there because the boundary is moving faster than the fluid below. One the other hand, the vorticity is positive on the sidewalls where the fluid is moving faster than the stationary walls. The vorticity does exhibit a slight asymmetry around $x = \frac{1}{2}$ due to advection.

**Midsection velocity.** The velocity at the midsection $x = \frac{1}{2}$ is calculated at $i = \frac{1}{2}N$ (note the need for $N$ to be even) as

$$u = \frac{\psi_{i\,j+1} - \psi_{i\,j}}{\Delta x}$$

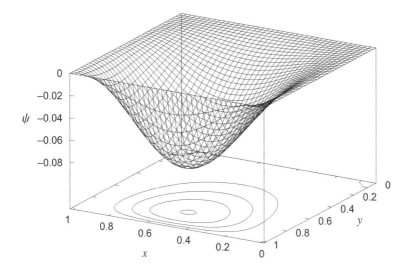

Figure 2.12 The steady-state streamfunction $\psi$; with $N = 40$ and $Re = 10$.

and plotted as a function of $y = (j + \frac{1}{2})\Delta x$ (which preserves the central differencing) in Figure 2.14. Note the good agreement from the different spatial resolutions. One can detect visually a small difference between the lowest resolution $N = 10$ + points and the higher resolutions. Note the velocity has the correct limits, $u \to 0$ as $y \to 0$ and $u \to 1$ as $y \to 1$.

**Steady force.** The force on the top plate (nondimensionalised by $\mu U_0$) is calculated as

$$F = \int_0^1 \left.\frac{\partial u}{\partial y}\right|_{y=1} dx \approx \sum_{i=0}^N \left.\frac{\partial^2 \psi}{\partial y^2}\right|_{j=N} \Delta x.$$

This will have second-order central differencing accuracy, because the shear stress vanishes at the two ends. With a first-order $O(\Delta x)$ error, we can evaluate the shear stress as

$$\left.\frac{\partial^2 \psi}{\partial y^2}\right|_{j=N} \approx \left.\frac{\partial^2 \psi}{\partial y^2}\right|_{j=N-1} = \frac{\psi_{iN} - 2\psi_{iN-1} + \psi_{i,N-2}}{\Delta x^2} + O(\Delta x).$$

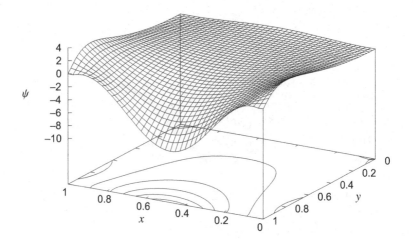

Figure 2.13  The steady-state vorticity $\omega$ with $N = 40$ and $Re = 10$.

Better is to extrapolate the second derivative linearly from the two interior points $N - 1$ and $N - 2$ onto the boundary $N$,

$$\left. \frac{\partial^2 \psi}{\partial y^2} \right|_{j=N} \approx 2 \left. \frac{\partial^2 \psi}{\partial y^2} \right|_{j=N-1} - \left. \frac{\partial^2 \psi}{\partial y^2} \right|_{j=N-2}$$

$$= \frac{2\psi_{iN} - 5\psi_{iN-1} + 4\psi_{i,N-2} - \psi_{i,N-3}}{\Delta x^2} + O(\Delta x^2).$$

One can check the last complicated formula gives the correct results for the second derivative taking in turn $\psi = 1$, $y$, $y^2$ and $y^3$ (with results 0, 0, 2 and 0).

Numerical results for the force $F$ are plotted as a function of the spatial resolution $\Delta x$ in Figure 2.15. The first simple expression for the shear stress clearly has a first-order $O(\Delta x)$ error, while the second more complex expression has at least a quadratic $O(\Delta x^2)$ error.

The final answer for the force is

$$F = 3.905 \pm 0.002 \quad \text{at } Re = 10.$$

The numerical results using the simple first simple expression for the force are not very accurate, the $N = 20$ being 20% wrong. However the clear linear dependence of the error on $\Delta x$ seen in Figure 2.15 can be used to extrapolate

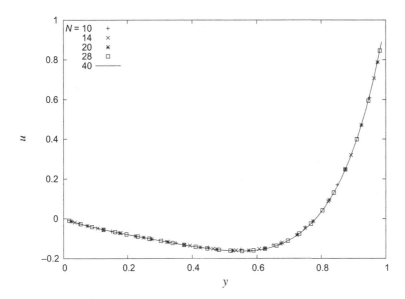

Figure 2.14 The steady-state horizontal velocity in the midsection $x = \frac{1}{2}$ at $Re = 10$, with $N = 10, 14, 20, 28$ and $40$.

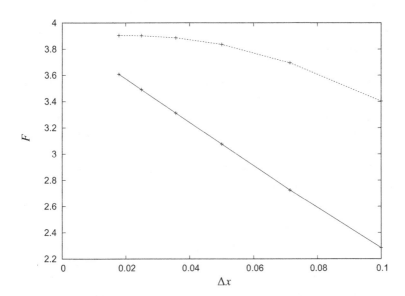

Figure 2.15 The steady-state force on the top plate at $Re = 10$ for $N = 10, 14, 20, 28, 40$ and $56$. The lower line is for the first $O(\Delta x)$ approximation, while the upper curve is the second $O(\Delta x^2)$ approximation.

results to $\Delta x = 0$ using

$$F_0 = \frac{F_2 \Delta x_1 - F_1 \Delta x_2}{\Delta x_1 - \Delta x_2}.$$

Applying this extrapolation to the first-order results for $N = 20$ and $N = 28$ results yields $F_0 = 3.906$.

**Force at early times.** Starting from rest, vorticity starts as a numerical version of a delta function against the moving top lid. It then diffuses down into a thin layer, of thickness $\sqrt{\nu t}$ dimensionally or $\sqrt{t/Re}$ in our nondimensionalisation. For early times the velocity profile near the top surface $y = 1$ will be

$$u(x, y, t) \sim U(x)\,\mathrm{erfc}\left(\frac{1 - y}{\sqrt{4t/Re}}\right),$$

which gives a force on the top lid changing in time

$$F(t) \sim \frac{1}{2}\sqrt{\frac{Re}{\pi t}}.$$

Note $\frac{1}{2\sqrt{\pi}} = 0.2821$. Figure 2.16 plots $F(t)/\sqrt{Re/t}$ as a function of time $t$ with different resolutions $N$. The first few times plotted are numerical nonsense. Extrapolating backwards to zero time, we find results considerably different from the expected $\frac{1}{2\sqrt{\pi}}$, 0.35 for $N = 40$, 0.33 for $N = 80$, 0.319 for $N = 160$ and 0.307 for $N = 320$. These results seem to converge to the correct answer with a large $0.4\Delta x^{1/2}$ error. This is an object lesson, in that the code which was good for the steady state is poor at resolving the singular behaviour of early times. If one needed to resolve accurately the early times, one would need a different approach, probably involving a stretch vertical coordinate whose stretching evolves in time.

In §15.3 about searching for singularities of physical problems, Figure 15.3 is a better plot of these early times.

# Exercises

It is now the time for students to solve the nonlinear flow using the formulation of streamfunction vorticity. MATLAB code is available on my website, but it would be better to write your own code.

**Exercise 2.10.1**   First set $Re = 10$, $N = 10$ and $\Delta t = 0.035$ and compute to $t = 0.525$. The plot of the streamfunction should show the numerical instability as in Figure 2.7. Decreasing $\Delta t$ to 0.03 nearly stabilises the

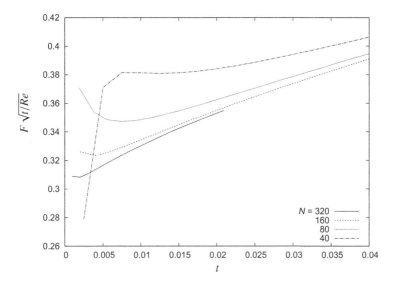

Figure 2.16 The force $F(t)$ at early times for $N = 40, 80, 160$ and 320.

result at this $t = 0.525$, while $\Delta t = 0.025$ works to $t = 3.0$. This value of $\Delta t = 0.025$ is the marginal value, but that is for large grids, and small grids are slightly more stable. It is best to work at $\Delta t = 0.2Re\Delta x^2$.

**Exercise 2.10.2**  The MATLAB code first calculates the boundary conditions to first order, and then extrapolates these results to apply the conditions to second order. Use this second-order code. Change to $Re = 10$, $N = 10$ and $\Delta t = 0.02$ and compute to $t = 1.0$. Find the value of $\omega(x = 0.5, y = 0.5, t = 1.0)$. Now decrease $\Delta t$ to 0.01, 0.005, 0.0025 and 0.001. Then with $N = 14$, try $\Delta t = 0.01$ (largest stable value), 0.005, 0.0025 and 0.001. Finally with $N = 20$, try $\Delta t = 0.005$ (largest stable value), 0.0025 and 0.001. Plot these results for $\psi(x = 0.5, y = 0.5, t = 1.0)$ as a function of $\Delta t$ to reproduce Figure 2.9. You could experiment by deleting the lines of the code that apply the boundary conditions at second order in order to reproduce Figure 2.8.

**Exercise 2.10.3**  Now find how long it takes the vorticity to attain a steady value within four significant figures. Compare your plots for the steady state of the streamfunction and the vorticity with those in Figures 2.12 and 2.13.

**Exercise 2.10.4**  Gather the results for different spatial resolutions $N$ for the steady horizontal velocity $u$ at the midsection $x = 0.5$, and plot on top of one another as in Figure 2.14.

**Exercise 2.10.5**    The code calculates the force on the top plate to second-order accuracy. Find the steady force for different spatial resolutions, $N = 10, 14, 20, 28$ and $40$. Plot the force as a function of the grid size $\Delta x$, to reproduce Figure 2.15.

**Exercise 2.10.6**    Change the top slip boundary condition from $u = \sin^2 \pi x$ to $u = 1$. Find the force on the top plate for various spatial resolutions, say $N = 10, 14, 20, 28$ and $40$. Show that the force diverges as the resolution increases as $F = 4.32 \ln(1/\Delta x) - 3.75$.

# 3

---

# Primitive variable formulation

## 3.1 Formulation

The streamfunction-vorticity formulation avoids the pressure problem, but is restricted to two-dimensional problems. We now tackle our two-dimensional problem of the driven cavity using the original variables of velocity $u, v$ and pressure $p$, which will enable us to tackle three-dimensional problems later.

The governing equations are

$$\frac{\partial u}{\partial t} = -u\frac{\partial u}{\partial x} - v\frac{\partial u}{\partial y} - \frac{\partial p}{\partial x} + \frac{1}{Re}\nabla^2 u,$$

$$\frac{\partial v}{\partial t} = -u\frac{\partial v}{\partial x} - v\frac{\partial v}{\partial y} - \frac{\partial p}{\partial y} + \frac{1}{Re}\nabla^2 v,$$

with pressure so that

$$\frac{\partial u}{\partial x} + \frac{\partial v}{\partial y} = 0.$$

The idea is that we use the first two equations to evolve the velocity forward in time, starting from a state of rest. The problem is how to find a pressure so that the third equation is satisfied.

## 3.2 Pressure equation

Now the incompressibility is true for all times, and so we can differentiate the incompressibility condition with respect to time,

$$\nabla \cdot \left(\frac{\partial \mathbf{u}}{\partial t}\right) = 0.$$

Taking the divergence of the momentum equation

$$\nabla \cdot \left( \frac{\partial \mathbf{u}}{\partial t} \right) = -\frac{\partial u_i}{\partial x_j} \frac{\partial u_j}{\partial x_i} - \mathbf{u} \cdot \nabla(\nabla \cdot \mathbf{u}) - \nabla^2 p + \frac{1}{Re} \nabla^2(\nabla \cdot \mathbf{u}).$$

Using $\nabla \cdot \mathbf{u} = 0$ in the above, we have an equation for the pressure

$$\nabla^2 p = -\frac{\partial u_i}{\partial x_j} \frac{\partial u_j}{\partial x_i}.$$

We need a boundary condition for the above equation for pressure. If we look at the $x$-component of the momentum equation,

$$\frac{\partial u}{\partial t} + u\frac{\partial u}{\partial x} + v\frac{\partial u}{\partial y} = -\frac{\partial p}{\partial x} + \frac{1}{Re}\left( \frac{\partial^2 u}{\partial x^2} + \frac{\partial^2 u}{\partial y^2} \right),$$

on the $x = 0$ sidewall boundary where $u = 0$ for all time $t$ and all vertical locations $y$, we obtain

$$0 + 0 + 0 = -\frac{\partial p}{\partial x} + \frac{1}{Re}\frac{\partial^2 u}{\partial x^2} + 0.$$

This condition generalises to all boundaries which have a vanishing normal component of velocity

$$\frac{\partial p}{\partial n} = \frac{1}{Re}\frac{\partial^2 u_n}{\partial n^2}.$$

Note that in specifying the normal derivative of the pressure on the boundary, the pressure will be determined up to an additive constant. Of course in our driven-cavity problem, a constant pressure drives no flow.

There is a minor complication that the pressure equation gives the 'source' of pressure in the interior while the boundary condition gives the 'flux' of pressure out over the boundary, and numerically these might not balance exactly, although any numerical algorithm will deliver an answer to such a contradictory problem.

Note that one cannot avoid the time-consuming solving of a Poisson problem. In the streamfunction-vorticity formulation, one has to solve a Poisson problem for the streamfunction for given vorticity. In the primitive variable formulation, one has to solve for the pressure. There are no free lunches.

## 3.3 Algorithm 1 with pressure equation

We can now construct an algorithm to integrate forward in time the incompressible Navier–Stokes equations in the primitive variable formulation. Starting from knowing the velocity field $\mathbf{u}(x, y, t)$ at time $t$, we can evaluate the

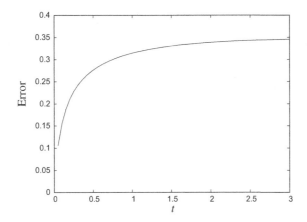

Figure 3.1 The evolution of the mean error in mass conservation, $|\nabla \cdot \mathbf{u}|$, for $N = 20$.

right-hand side forcing of the Poisson equation for the pressure along with its normal gradient boundary condition. Solving this Poisson problem by the numerical methods used in §2.3 to find the streamfunction from the vorticity, we can find the pressure field $p(x, y, t)$ at time $t$. Hence we can evaluate the pressure gradient to use in the momentum equation, so finding the time derivative of the velocity field at time $t$, $\partial \mathbf{u}/\partial t$. Thus we can make the step forward in time from $t$ to $t + \Delta t$.

This algorithm based on the pressure equation DOES NOT WORK.

In our driven-cavity problem, the algorithm goes to a strange steady state in which the horizontal velocity $u$ is erroneously positive everywhere in the box, satisfying the boundary conditions and decaying away from the top moving boundary. The second spatial derivative of this strange horizontal velocity gives a positive pressure gradient on the two side walls. This drives a pressure with a minimum at the left end of the moving top boundary and a maximum at the right end. The pressure minimum sucks in a positive vertical velocity $v$ towards the moving lid on the left side, while the maximum pushes away a negative vertical velocity on the right side. Note that the recirculation by negative horizontal velocity $u$ in the lower half of the box does not happen, i.e. the strange steady solution does not conserve mass.

The evolution of the failure to conserve mass, evaluated as the mean over the grid of the absolute value of the divergence of the velocity, $|\nabla \cdot \mathbf{u}|$, is plotted in Figure 3.1. Halving the time-step does not change the curve. Increasing the spatial resolution from $N = 20$ to $N = 30$ *increases* the error by 10%.

The algorithm based on the pressure equation fails because it assumes that

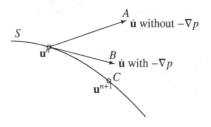

Figure 3.2 Incompressibility as a constraint.

the velocity $\mathbf{u}$ at time $t$ is divergence-free, and using this assumption finds a time-derivative $\partial\mathbf{u}/\partial t$ which is divergence-free. Note the time-derivative will not be divergence-free if the starting velocity is not. Note that any numerical error in the divergence made when stepping forward in time is not corrected by the algorithm. Hence numerical errors accumulate and come to dominate the final steady state.

## 3.4 Incompressibility as a constraint: split time-step

In the space of all vector fields $\mathbf{u}(x, y, t)$, the incompressibility condition constrains the solution $\mathbf{u}$ of the Navier–Stokes equation to the surface $\nabla \cdot \mathbf{u} = 0$. The role of the pressure gradient $-\nabla p$ in the momentum equation is to keep the solution on this surface. Thus the pressure *projects* out the component of $\partial\mathbf{u}/\partial t$ which is out of the constrained surface.

Due to the accumulation of small numerical errors, the numerical solution can drift slowly away from the constraint surface $\nabla \cdot \mathbf{u} = 0$. To avoid this drift, we project the velocity *at the end of the time-step* back onto the constraint surface, instead of just projecting out the component of $\dot{\mathbf{u}}$ out of the surface. This idea is illustrated in Figure 3.2. The solution $\mathbf{u}^n$ at time $t = n\Delta t$ starts in the constraint surface $S$. Omitting the pressure gradient from the momentum equation, $\dot{\mathbf{u}}$ would take the solution to point $A$ outside the constraint surface. Including the pressure gradient makes $\dot{\mathbf{u}}$ tangential to the surface, but takes the solution to point $B$ a little above it. Our aim is to do better and to arrive at the end of the time-step with $\mathbf{u}^{n+1}$ at point $C$.

The projection back onto the constraint surface is achieved with a split time-step approach. First one steps forward in time from $\mathbf{u}^n$ to an intermediate $\mathbf{u}^*$ (point $A$ in Figure 3.2) using the momentum equation without the pressure gradient term

$$\mathbf{u}^* = \mathbf{u}^n + \Delta t \left( -\mathbf{u}^n \cdot \nabla \mathbf{u}^n + \frac{1}{Re} \nabla^2 \mathbf{u}^n \right).$$

Then the projection part of the step is made adding in the effect of the pressure gradient with

$$\mathbf{u}^{n+1} = \mathbf{u}^* - \Delta t \nabla p,$$

where the pressure field $p(x, y)$ is chosen so that one ends up on the constraint surface at point $C$ in Figure 3.2, i.e. $\nabla \cdot \mathbf{u}^{n+1} = 0$. Hence the pressure must satisfy

$$\nabla^2 p = \frac{1}{\Delta t} \nabla \cdot \mathbf{u}^*.$$

This split time-step approach corrects any errors in satisfying the constraint in the past, $\nabla \cdot \mathbf{u}^n \neq 0$, as well as errors in the first part of the split step.

The first part of the split step to $\mathbf{u}^*$ is straightforward, but solving for the pressure has some complications. For the first part we have

$$u^*_{ij} = u^n_{ij}$$
$$+ \Delta t \left( -u^n_{ij} \frac{u^n_{i+1\,j} - u^n_{i-1\,j}}{2\Delta x} - v^n_{ij} \frac{u^n_{i\,j+1} - u^n_{i\,j+1}}{2\Delta x} \right)$$
$$+ \frac{\Delta t}{Re\Delta x^2} \begin{pmatrix} & 1 & \\ 1 & -4 & 1 \\ & 1 & \end{pmatrix} u^n_{ij},$$

and a similar expression for $v^*_{ij}$. For the projection step, solving the above Poisson equation for pressure by

$$\frac{\Delta t}{\Delta x^2} \begin{pmatrix} & 1 & \\ 1 & -4 & 1 \\ & 1 & \end{pmatrix} p_{ij} = \frac{u^*_{i+1\,j} - u^*_{i-1\,j}}{2\Delta x} + \frac{v^*_{i\,j+1} - v^*_{i\,j-1}}{2\Delta x}$$

does not quite give the desired $\nabla \cdot \mathbf{u}^{n+1} = 0$, but has a small error which tends to zero as $\Delta x \to 0$. It is a moot point whether one should be bothered about a small error in satisfying the incompressibility when there is a similar small error in satisfying the momentum equation.

## 3.5 Algorithm 2 by projection – spurious pressure modes

To ensure $\nabla \cdot \mathbf{u}^{n+1}$ vanishes exactly, we need to be a little more careful evaluating this divergence on the grid. Now with our central differencing

$$\left.\frac{\partial u^{n+1}}{\partial x}\right|_{ij} = \frac{u_{i+1\,j}^{n+1} - u_{i-1\,j}^{n+1}}{2\Delta x}$$

$$= \frac{\left(u_{i+1\,j}^* - \Delta t \dfrac{p_{i+2\,j} - p_{ij}}{2\Delta x}\right) - \left(u_{i-1\,j}^* - \Delta t \dfrac{p_{ij} - p_{i-2\,j}}{2\Delta x}\right)}{2\Delta x},$$

and similarly for $\partial v^{n+1}/\partial y$. Hence the pressure must satisfy at internal points $i = 1, \ldots, N-1$ and $j = 1, \ldots, N-1$

$$\frac{\Delta t}{4\Delta x^2} \begin{pmatrix} & & 1 & & \\ & & 0 & & \\ 1 & 0 & -4 & 0 & 1 \\ & & 0 & & \\ & & 1 & & \end{pmatrix} p_{ij} = \left(\frac{u_{i+1\,j}^* - u_{i-1\,j}^*}{2\Delta x} + \frac{v_{i\,j+1}^* - v_{i\,j-1}^*}{2\Delta x}\right).$$

The difference between this equation and the one at the end of the previous section stems from the feature of numerical differentiation that $f'' \neq (f')'$.

At internal points adjacent to the boundary, the preceding expression above for $\nabla \cdot \mathbf{u}^{n+1}$ calls for the value on the boundary of the normal component of the velocity, which is zero for our boundary conditions. Because this normal component vanishes in $\mathbf{u}^{n+1}$ and $\mathbf{u}^*$, we must set the pressure at the one point outside the boundary equal to its value one point inside, e.g. on the left-hand boundary $p_{-1\,j} = p_{1,j}$. With this boundary condition, which is not the true boundary condition on the pressure found in §3.2 (see also discussion in the penultimate paragraph of §3.6), we can solve the Poisson problem for the pressure.

The solution for the pressure is shown in Figure 3.3. We see very undesirable oscillations in the solution, due to a 'spurious' pressure mode. The wide numerical molecule involving $i-2$, $i$ and $i+2$ makes the pressure solution on the even $i$ decouple from the solution on the odd $i$. Note that a pressure solution $p_{ij} = (-1)^{i+j}$ produces no change from $\mathbf{u}^*$ to $\mathbf{u}^{n+1}$. We therefore need a more compact numerical molecule, where such an alternating pressure field would have an effect on the velocity.

While the spurious pressure mode does not act in the projection from $\mathbf{u}^*$ to $\mathbf{u}^{n+1}$, the velocity field resulting from algorithm 2 does have some small oscillations, as can be seen in Figure 3.4. These small oscillations come from the small differences in the gradients of the pressure evaluated from the decoupled odd- and even-node pressure solutions. There are tricks available to

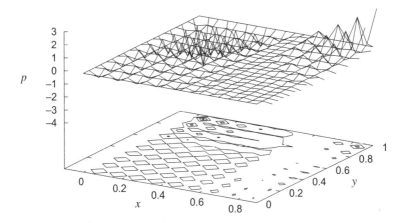

Figure 3.3 The pressure field from algorithm 2 and $N = 20$ showing a spurious pressure mode.

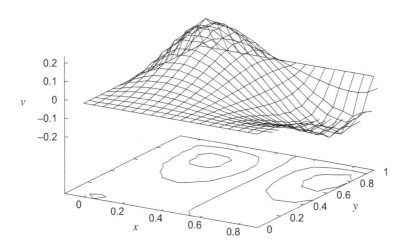

Figure 3.4 The vertical velocity $v$ from algorithm 2 and $N = 20$ showing small oscillations.

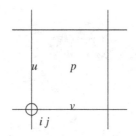

Figure 3.5 The staggered grid, with $u_{i\,j+\frac{1}{2}}$, $v_{i+\frac{1}{2}\,j}$ and $p_{i+\frac{1}{2}\,j+\frac{1}{2}}$.

pull the decoupled pressure solutions together, essentially by damping spurious short wavelength disturbances, but these tricks sacrifice exactly satisfying the incompressibility constraint.

## 3.6 Algorithm 3 with a staggered grid

The spurious pressure mode in algorithm 2 resulted from the wide numerical molecule in the pressure equation which decoupled the odd and even points on the grid. To make a compact numerical molecule for the pressure which avoids this problem, we need the new idea of a *staggered grid*, where different variables are held at different locations on the grid; see Figure 3.5. The horizontal velocity $u$ is held in the middle of the vertical sides of a grid block, i.e., at locations $i + \frac{1}{2}\,j$, and the vertical velocity is held in the middle of the horizontal sides of a grid block, i.e., at locations $i\,j + \frac{1}{2}$. These locations describe the flux of volume into the block. The pressure is held in the middle of the grid block at location $i + \frac{1}{2}\,j + \frac{1}{2}$. When coding this algorithm, the variables are stored with indices rounded down to an integer.

The sides of the box are made to coincide with the grid lines, i.e., the sides are $i = 0$ and $i = N$, the bottom is $j = 0$ and the top $j = N$. With this choice, the components of velocity involved in the boundary condition of no normal mass flux are held on the boundary; see Figure 3.6. Hence, this boundary condition is

$$u_{0\,j+\frac{1}{2}} = u_{N\,j+\frac{1}{2}} = 0 \quad \text{for } j = 0 \to N - 1,$$

$$v_{i+\frac{1}{2}\,0} = v_{i+\frac{1}{2}\,N} = 0 \quad \text{for } i = 0 \to N - 1.$$

The tangential component of velocity is held half a grid block away from the surface. To impose the no-slip boundary conditions on the boundaries, one sets the tangential velocity at the half-grid block just outside the boundary; see

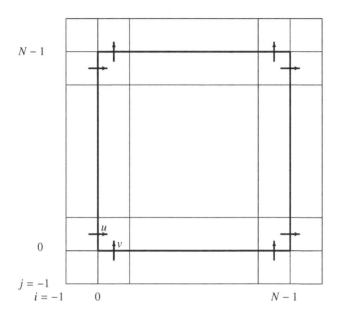

Figure 3.6 The sides of the box, in darker lines, are chosen to coincide with the velocity components corresponding to the no-normal-flux boundary condition.

Figure 3.7, such that linear interpolation onto the boundary takes the prescribed value, i.e.

$$v_{-\frac{1}{2}j} = -v_{\frac{1}{2}j} \quad \text{and} \quad v_{N+\frac{1}{2}j} = -v_{N-\frac{1}{2}j} \quad \text{for } j = 1 \rightarrow N - 1,$$

$$u_{i-\frac{1}{2}} = -u_{i\frac{1}{2}} \quad \text{and} \quad u_{iN+\frac{1}{2}} = 2\sin^2(i * \Delta x) - u_{iN-\frac{1}{2}} \quad \text{for } i = 1 \rightarrow N - 1.$$

The momentum equation is satisfied with a split time-step, first advancing from $\mathbf{u}^n$ to $\mathbf{u}^*$ omitting the pressure gradient, and then making the projection from $\mathbf{u}^*$ to an exactly incompressible $\mathbf{u}^{n+1}$ through the action of a suitable pressure gradient. For the first part, we evaluate the time derivative of the components of velocity at the locations they are held using central spatial

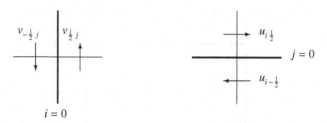

Figure 3.7 To satisfy the zero tangential velocity boundary condition, the tangential velocity at the ghost point just outside the boundary is set to the opposite of the value just inside the boundary.

differencing. Thus

$$
u^*_{i\,j+\frac{1}{2}} = u^n_{i\,j+\frac{1}{2}}
$$

$$
- \Delta t\, u^n_{i\,j+\frac{1}{2}} \frac{u^n_{i+1\,j+\frac{1}{2}} - u^n_{i-1\,j+\frac{1}{2}}}{2\Delta x}
$$

$$
- \Delta t \frac{1}{4} \left( v^n_{i+\frac{1}{2}\,j} + v^n_{i-\frac{1}{2}\,j} + v^n_{i+\frac{1}{2}\,j+1} + v^n_{i-\frac{1}{2}\,j+1} \right) \frac{u^n_{i\,j+\frac{3}{2}} - u^n_{i\,j-\frac{1}{2}}}{2\Delta x}
$$

$$
+ \frac{\Delta t}{Re\Delta x^2} \begin{bmatrix} & 1 & \\ 1 & -4 & 1 \\ & 1 & \end{bmatrix} u^n_{i\,j+\frac{1}{2}},
$$

and a similar expression for $v^*_{i+\frac{1}{2}\,j}$. Note in the expression above the spatial derivatives are compact and the only complication is the need to average $v$ over the four values which surround the point $i\,j+\frac{1}{2}$ on the staggered grid.

The incompressibility condition on $\mathbf{u}^{n+1}$ is imposed at the location $i+\frac{1}{2}\,j+\frac{1}{2}$ where pressure is held on the staggered grid. The resulting equation for pressure is

$$
\frac{\Delta t}{\Delta x^2} \begin{bmatrix} & 1 & \\ 1 & -4 & 1 \\ & 1 & \end{bmatrix} P_{i+\frac{1}{2}\,j+\frac{1}{2}} = \frac{u^*_{i+1\,j+\frac{1}{2}} - u^*_{i\,j+\frac{1}{2}}}{\Delta x} + \frac{v^*_{i+\frac{1}{2}\,j+1} - v^*_{i+\frac{1}{2}\,j}}{\Delta x}.
$$

The pressure boundary condition from §3.2 applied at second order becomes on the left boundary

$$p_{-\frac{1}{2}\,j+\frac{1}{2}} = p_{\frac{1}{2}\,j+\frac{1}{2}} + \frac{1}{Re}\frac{-u_{3\,j+\frac{1}{2}} + 4u_{2,j+\frac{1}{2}} - 5u_{1\,j+\frac{1}{2}} + 2u_{0\,j+\frac{1}{2}}}{\Delta x},$$

and similar expressions on the other boundaries.

Finally the projection part of the time-step is

$$u^{n+1}_{i\,j+\frac{1}{2}} = u^{*}_{i\,j+\frac{1}{2}} - \Delta t\frac{p_{i+\frac{1}{2}\,j+\frac{1}{2}} - p_{i-\frac{1}{2}\,j+\frac{1}{2}}}{\Delta x},$$

at the internal points, and a similar expression for $v^{n+1}_{i+\frac{1}{2}\,j}$.

The above algorithm produces an exactly solenoidal $\mathbf{u}^{n+1}$ except in the grid blocks adjacent to the boundary. The problem comes from the pressure boundary condition which says that there is a pressure gradient on the boundary. On the other hand, by setting the normal component of the velocity in $\mathbf{u}^{*}$ and $\mathbf{u}^{n+1}$ equal to zero on the boundary, we are effectively saying in the projection step that the normal pressure gradient is zero effectively because the projection step was not applied to the boundary points. To cure this problem we need to apply the pressure projection step to the boundary as well as the interior points. This would mean changing the normal component of velocity from $\mathbf{u}^{*}$ to $\mathbf{u}^{n+1}$. In order to finish the time-step with a vanishing normal component, we need to advance from zero the normal component in $\mathbf{u}^{*}$ by the amount it will be projected back to zero in $\mathbf{u}^{n+1}$. This makes sense because the momentum equation on the boundary was found in §3.2 to be

$$\frac{\partial u_n}{\partial t} = -\frac{\partial p}{\partial n} + \frac{1}{Re}\frac{\partial^2 u_n}{\partial n^2}.$$

In the first part of the split time-step we are supposed to advance $\mathbf{u}^n$ to $\mathbf{u}^{*}$ using the momentum equation with the pressure gradient term omitted. Doing this on the boundary we will have a nonzero normal velocity on the boundary in $\mathbf{u}^{*}$ through the viscous term. But the pressure gradient exactly balances, this so will return it to zero in $\mathbf{u}^{n+1}$.

There are many numerical codes and software packages which erroneously set the normal gradient of the pressure to zero on the boundaries, and also set the normal component of $\mathbf{u}^{*}$ to zero. It is clearly erroneous because nearly all analytic solutions of viscous flow will show that the normal gradient of the pressure is nonzero. Applying this condition of zero normal pressure gradient seems not to affect the pressure variations within the interior. The wrong gradient at the boundary therefore just introduces an $O(\Delta x)$ error in the value of the pressure on the boundary.

In the later §7.10, we describe the Pressure-Update method of time-stepping,

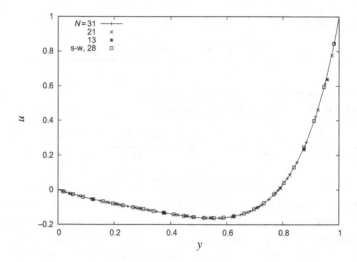

Figure 3.8 The steady-state horizontal velocity in the midsection $x = \frac{1}{2}$ at $Re = 10$ from algorithm 3 with a staggered grid and $N = 14, 20$ and $30$. Also plotted is the result from the streamfunction-vorticity formulation at $N = 28$.

which avoids some of the problems of the Pressure Projection method of this chapter.

## 3.7 Results from algorithm 3

As with the streamfunction-vorticity formulation, one first needs to test the code, to ensure that the time-stepping is numerically stable, that the time evolution is first-order accurate in $\Delta t$ and second-order accurate in $\Delta x$. These tests were of course made, but the evidence is omitted. Similarly one has to check the time taken to achieve a steady state.

**Midsection velocity.** The horizontal velocity in the midsection $x = \frac{1}{2}$ is evaluated from $u_{i\,j+\frac{1}{2}}$ at $i = \frac{1}{2}N$ (need even $N$). This is plotted as a function of $y = (j + \frac{1}{2})\Delta x$ in Figure 3.8. Note the good agreement again between different spatial resolutions, and also the agreement with the previous results from the streamfunction-vorticity formulation. Confidence in a code increases enormously when it produces results coinciding with results from a totally different code.

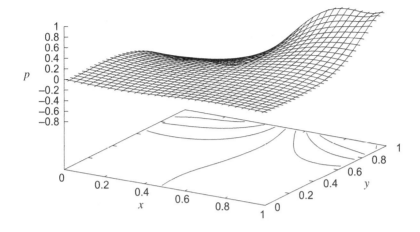

Figure 3.9 The pressure field from algorithm 3 with a staggered grid at $N = 30$.

**Pressure.** The new result from the primitive variable formulation is the pressure field, and this is plotted in Figure 3.9 for $N = 30$. There is a pressure minimum in the top-left corner, which sucks fluid vertically towards the beginning of the top moving plate. There is a pressure maximum in the top-right corner, which blows fluid vertically away from the end of the top moving plate. In the lower part of the box, the pressure gradient from right to left drives the horizontal return flow from right to left. Thus the pressure ensures the fluid conserves mass and does not accumulate anywhere.

**Force.** The viscous stress on the top plate is easy to evaluate with second-order accuracy on the staggered grid

$$F = \sum_{i=1}^{N-1} \frac{u_{i\,N+\frac{1}{2}} - u_{i\,N-\frac{1}{2}}}{\Delta x} \Delta x + O(\Delta x^2).$$

There is a similar expression for the viscous force $F_b$ on the bottom. Now that the pressure is available, we can also evaluate the pressure force on the sidewalls, averaging the values on the staggered grid either side of the sidewall,

$$F_p = \sum_{j=0}^{N-1} \frac{1}{2} \left( -p_{-\frac{1}{2}\,j+\frac{1}{2}} - p_{\frac{1}{2}\,j+\frac{1}{2}} + p_{N-\frac{1}{2}\,j+\frac{1}{2}} + p_{N+\frac{1}{2}\,j+\frac{1}{2}} \right) + O(\Delta x^2).$$

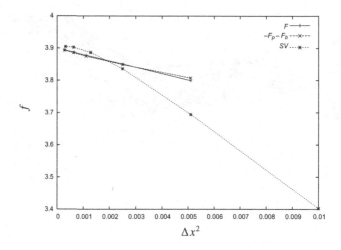

$\Delta x^2$

Figure 3.10 The steady-state force at $Re = 10$ as a function of the spatial resolution, from algorithm 3 calculating the viscous force on the top $F$ and the pressure force on the sides $F_p$ and viscous force on the bottom $F_b$. Also plotted are the results from the streamfunction-vorticity formulation using the second-order boundary condition.

Figure 3.10 gives the viscous force on the top $F$ as a function of the spatial resolution. Also plotted are the sum of the pressure forces on the sidewalls and the viscous force on the bottom $-F_p - F_b$. These almost coincide, which is reassuring (but see §4.5). For comparison, the figure also has the result from §2.10 for the streamfunction-vorticity formulation using the second-order accurate boundary condition. There is a clear quadratic dependence on the spatial resolution $\Delta x$ for the primitive variable formulation. Extrapolating to zero, the force on the top is equal to the force on on the sides and bottom, with a value $3.8998 \pm 0.0002$. The viscous force on the bottom is relatively small at $-0.254$.

## Exercises

Writing the code for the primitive variable formulation is a little more complicated. Matlab code is available on my website for the version of algorithm 3 with a staggered grid.

**Exercise 3.7.1**  First try some tests, as suggested for the streamfunction-vorticity formulation, to check the stability of time-stepping, the second-order accuracy and the time to reach the steady state.

**Exercise 3.7.2**    Gather the results for different spatial resolutions $N$ for the steady horizontal velocity at the midsection $x = 0.5$, and plot these results on top of one another as in Figure 3.8. Include a result from the streamfunction-vorticity formulation.

**Exercise 3.7.3**    The code calculates the force to second-order accuracy on the top plate, on the sides and on the bottom. Reproduce Figure 3.10.

**Exercise 3.7.4**    Experiment with different geometries of the cavity, say twice as deep as wide. The optimal parameter for the successive overrelaxation (SOR) will change with the geometry, but this should not affect the calculation of the steady state. The force on the bottom plate should decrease rapidly with deeper cavities, while the force on the top plate should decrease only slightly. On the other hand for shallower cavities, the force on the top plate increases as the height decreases.

**Exercise 3.7.5**    Experiment with different Reynolds numbers, say $Re = 10$, say 5, 2 and 20. The code reduces the time-step as $Re$ is decreased. That will increase the run-time, although the steady state should be reached earlier. At larger values of $Re$ the time-step may be limited by the CFL condition if the spatial resolution is not increased. The forces change very little in $Re < 10$, i.e. are effectively in the low Reynolds number limit. For higher Reynolds numbers, $Re > 50$, the forces increase as the top boundary layer begins to thin. Note that at these higher Reynolds numbers the steady state is achieved later, say by $t = 10$.

# PART II

## GENERALITIES

# 4

---

# Finite differences

Part I tackled a simple problem by simple methods. Along the way various general issues were revealed, such as discretisation, time-stepping and large sparse linear equations. In Part II we look in greater detail at these general issues. There are three ways to turn continuous partial differential equations into a finite discrete problem. In this chapter we consider again finite differences, to be followed by subsequent chapters on finite elements and spectral methods.

In finite differencing, we hold the unknown functions at a discrete number of points on a grid. Normally the points are equispaced with a separation of $\Delta x$.

## 4.1 Higher orders

### 4.1.1 Central differencing

For equispaced points, we can use central differencing to obtain a second-order estimate for derivatives. We have already used expressions for the first and second derivatives

$$f_i' = \frac{f_{i+1} - f_{i-1}}{2\Delta x} + O(\Delta x^2),$$
$$f_i'' = \frac{f_{i+1} - 2f_i + f_{i-1}}{\Delta x^2} + O(\Delta x^2).$$

We can use these to find higher-order derivatives, e.g.

$$f_i''' = \frac{f_{i+1}'' - f_{i-1}''}{2\Delta x} + O(\Delta x^2)$$

$$= \frac{f_{i+2} - 2f_{i+1} + 2f_{i-1} - f_{i-2}}{2\Delta x^3} + O(\Delta x^2)$$

$$f_i'''' = \frac{f_{i+1}'' - 2f_i'' + f_{i-1}''}{\Delta x^2} + O(\Delta x^2)$$

$$= \frac{f_{i+2} - 4f_{i+1} + 6f_i - 4f_{i-1} + f_{i-2}}{\Delta x^4} + O(\Delta x^2),$$

and so on. The coefficients in the even derivatives are the binomial coefficients in Pascal's triangle. The coefficients in the odd derivatives are the difference between the binomial coefficients with a positive and negative shift.

The local errors can be found using a Taylor series

$$f_{i+1} = f(x = i\Delta x + \Delta x)$$

$$= f_i + \Delta x f_i' + \tfrac{1}{2}\Delta x^2 f_i'' + \tfrac{1}{6}\Delta x^3 f_i''' + \tfrac{1}{24}\Delta x^4 f_i'''' + \cdots.$$

Then for the difference in the first derivative above

$$f_{i+1} - f_{i-1} = 2\Delta x f_i' + \tfrac{1}{3}\Delta x^3 f_i''' + O(\Delta x^5).$$

Thus we have an expression for the local error

$$f_i' = \frac{f_{i+1} - f_{i-1}}{2\Delta x} - \tfrac{1}{6}\Delta x^2 f_i''' + O(\Delta x^4).$$

But we have above a second-order accurate expression for the third derivative $f_i'''$. Substituting this, we can obtain a fourth-order accurate expression for the first derivative,

$$f_i' = \frac{-\tfrac{1}{12}f_{i+2} + \tfrac{2}{3}f_{i+1} - \tfrac{2}{3}f_{i-1} + \tfrac{1}{12}f_{i-2}}{\Delta x} + O(\Delta x^4).$$

One should check this expression, that $f = 1, x, x^2, x^3, x^4$ all give the correct result $(0, 1, 0, 0, 0)$.

Similarly, we can obtain a fourth-order accurate expression for the second derivative

$$f_i'' = \frac{-\tfrac{1}{12}f_{i+2} + \tfrac{4}{3}f_{i+1} - \tfrac{5}{2}f_i + \tfrac{4}{3}f_{i-1} - \tfrac{1}{12}f_{i-2}}{\Delta x^2} + O(\Delta x^4).$$

### 4.1.2 One-sided differencing

At boundaries it is not possible to use central differencing, because one would need information about the function outside the domain. Hence there is a need to use information from points on just one side. Still with equispaced points we have

$$f_0' = \frac{f_1 - f_0}{\Delta x} + O(\Delta x),$$

$$f_0'' = \frac{f_2 - 2f_1 + f_0}{\Delta x^2} + O(\Delta x),$$

$$f_0''' = \frac{f_3 - 3f_2 + 3f_1 - f_0}{\Delta x^3} + O(\Delta x).$$

We can again find an expression for the local error using Taylor series. Thus for the one-sided first derivative above

$$f_1 - f_0 = \Delta x f_0' + \tfrac{1}{2}\Delta x^2 f_0'' + O(\Delta x^3).$$

Using the first-order expression above for the second derivative, we obtain the improved second-order accurate expression for the one-sided first derivative

$$f_0' = \frac{-\tfrac{1}{2}f_2 + 2f_1 - \tfrac{3}{2}f_0}{\Delta x} + O(\Delta x^2).$$

Similarly one can find

$$f_0'' = \frac{-f_3 + 4f_2 - 5f_1 + 2f_0}{\Delta x^2} + O(\Delta x^2).$$

These expressions were used in the driven-cavity problem when extracting information about derivatives on the boundary to second-order accuracy.

### 4.1.3 Nonequispaced points

To find an expression for the $k$th derivative $f^{(k)}(x_0)$ with accuracy $O(\Delta x^l)$ one fits a polynomial of degree $k + l$ through $k + l + 1$ points $x_0 + \Delta x_i$,

$$f(x_0 + \Delta x_i) = a_0 + a_1 \Delta x_i + a_2 \Delta x_i^2 + \cdots a_{k+l} \Delta x_i^{k+l}.$$

One solves for the polynomial coefficients $a_j$, typically using a computer-algebra package such as MAPLE. Finally the desired $k$th derivative is

$$f^{(k)}(x_0) = a_k \, k!.$$

Central differencing on equispaced points is to be preferred because it delivers one degree higher accuracy than the same number of points would give if they were not equispaced or the differencing were not central.

It is not normally worth going to higher accuracy than the lowest approximation or its first improvement. At higher accuracy, the expressions use information from a wider spread of points. It is known that fitting high-order polynomials through fixed points soon generates polynomials which differ greatly from the function in between the fitting points. A much better approach would be to use splines see §11.1.1.

## 4.2 Compact fourth-order Poisson solver

### 4.2.1 One-dimensional version

First we consider

$$\frac{d^2\phi}{dx^2} = \rho.$$

Using fourth-order accurate central differencing, we have the discretised form

$$-\tfrac{1}{12}\phi_{i+2} + \tfrac{4}{3}\phi_{i+1} - \tfrac{5}{2}\phi_i + \tfrac{4}{3}\phi_{i-1} - \tfrac{1}{12}\phi_{i-2} = \Delta x^2 \rho_i.$$

This wide numerical molecule gives problems next to the boundaries because it requests values of $\phi$ outside the boundary. The large width also tends to have large coefficients in the $O(\Delta x^4)$ error.

To find a more compact approach, we start by considering again the error in the second-order approximation to the second derivative

$$\frac{\phi_{i+1} - 2\phi_i + \phi_{i-1}}{\Delta x^2} = \phi_i'' + \tfrac{1}{6}\Delta x^2 \phi_i'''' + O(\Delta x^4).$$

Now $\phi_i'' = \rho_i$ and so

$$\phi_i'''' = \rho_i'' = \frac{\rho_{i+1} - 2\rho_i + \rho_{i+1}}{\Delta x^2} + O(\Delta x^2).$$

Hence

$$\frac{\phi_{i+1} - 2\phi_i + \phi_{i-1}}{\Delta x^2} = \tfrac{1}{6}\rho_{i+1} + \tfrac{2}{3}\rho_i + \tfrac{1}{6}\rho_{i-1} + O(\Delta x^4).$$

This expression gives an algorithm for the solution of the Poisson problem in one dimension with a fourth-order accurate answer.

### 4.2.2 Two dimensions

We use the same idea now in two dimensions. Thus we use

$$\nabla^2 \rho = \nabla^2 \nabla^2 \phi = \frac{\partial^4 \phi}{\partial x^4} + 2\frac{\partial^4 \phi}{\partial x^2 \partial y^2} + \frac{\partial^4 \phi}{\partial y^4}.$$

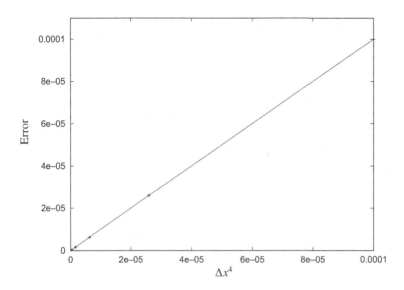

Figure 4.1 Test for fourth-order error, with $N = 10, 14, 20, 40$ and $56$.

There is a little complication that our standard numerical molecule for the Laplacian does not have an error of the above form, but

$$\begin{pmatrix} & 1 & \\ 1 & -4 & 1 \\ & 1 & \end{pmatrix}\phi = \Delta x^2\nabla^2\phi + \tfrac{1}{12}\Delta x^4\left(\frac{\partial^4\phi}{\partial x^4} + \frac{\partial^4\phi}{\partial y^4}\right).$$

Fortunately an alternative form of the Laplacian, one rotated by $\tfrac{1}{4}\pi$, has a different combination of fourth derivatives

$$\begin{pmatrix} \tfrac{1}{2} & & \tfrac{1}{2} \\ & -2 & \\ \tfrac{1}{2} & & \tfrac{1}{2} \end{pmatrix}\phi = \Delta x^2\nabla^2\phi + \tfrac{1}{12}\Delta x^4\left(\frac{\partial^4\phi}{\partial x^4} + 6\frac{\partial^4\phi}{\partial x^2\partial y^2} + \frac{\partial^4\phi}{\partial y^4}\right).$$

Then taking $\tfrac{2}{3}$ of the first and $\tfrac{1}{3}$ of the second expression, we have an error which is the Laplacian of $\rho$. This gives the compact fourth-order accurate algorithm

$$\frac{1}{\Delta x^2}\begin{pmatrix} \tfrac{1}{6} & \tfrac{2}{3} & \tfrac{1}{6} \\ \tfrac{2}{3} & -\tfrac{10}{3} & \tfrac{2}{3} \\ \tfrac{1}{6} & \tfrac{2}{3} & \tfrac{1}{6} \end{pmatrix}\phi = \begin{pmatrix} 0 & \tfrac{1}{12} & 0 \\ \tfrac{1}{12} & \tfrac{2}{3} & \tfrac{1}{12} \\ 0 & \tfrac{1}{12} & 0 \end{pmatrix}\rho + O(\Delta x^4).$$

As in §2.4, this algorithm is tested by the same analytic solution

$$\rho = 2\pi^2 \sin \pi x \sin \pi y \quad \text{and} \quad \phi = -\sin \pi x \sin \pi y.$$

In Figure 4.1 we see an error decreasing as $0.27\Delta x^4$. Thus at $N = 20$, this fourth-order algorithm gives an error of $2\,10^{-6}$ compared with the earlier second-order algorithm which gave an error of $2\,10^{-3}$ at this resolution.

The same trick of examining the leading error and using the governing equation applied twice can be used to derive the compact Crandall fourth-order algorithm for the diffusion equation $u_t = u_{xx}$:

$$u_i^{n+1} + \left(\frac{1}{12} - \frac{\Delta t}{2\Delta x^2}\right)\left(u_{i+1}^{n+1} - 2u_i^{n+1} + u_{i-1}^{n+1}\right)$$

$$= u_i^n + \left(\frac{1}{12} + \frac{\Delta t}{2\Delta x^2}\right)\left(u_{i+1}^n - 2u_i^n + u_{i-1}^n\right).$$

## 4.3  Upwinding

The advection term $\mathbf{u}\cdot\nabla\phi$ propagates information in the direction of $\mathbf{u}$. This is violated by the central difference approximation

$$u_i\frac{\phi_{i+1} - \phi_{i-1}}{2\Delta x},$$

where when $u_i > 0$ downstream information $\phi_{i+1}$ will influence the time evolution of $\phi_i$.

One can correct this erroneous flow of information by evaluating the spatial derivative using only upstream data, i.e. using the earlier one-sided derivatives

$$u\frac{\partial\phi}{\partial x} = \begin{cases} u_i\dfrac{\phi_i - \phi_{i-1}}{\Delta x} & \text{if } u_i > 0, \\[2ex] u_i\dfrac{\phi_{i+1} - \phi_i}{\Delta x} & \text{if } u_i < 0, \end{cases}$$

and a similar expression for $v\partial\phi/\partial y$. These one-sided derivatives are only first-order accurate compared with the second-order accurate central differencing. One can of course use extra data points for a second-order one-sided derivative

$$u\frac{\partial\phi}{\partial x} = \begin{cases} u_i\dfrac{3\phi_i - 2\phi_{i-1} + \phi_{i-2}}{2\Delta x} & \text{if } u_i > 0, \\[2ex] u_i\dfrac{-\phi_{i+2} + 2\phi_{i+1} - 3\phi_i}{2\Delta x} & \text{if } u_i < 0. \end{cases}$$

This, along with the similar expression for $v\partial\phi/\partial y$, has a rather wide numerical molecule. A more compact molecule which is not always strictly upwinding is for $u > 0$ and $v > 0$

$$\frac{u}{\Delta x}\begin{pmatrix} 0 & 0 & 0 \\ -1 & 1 & 0 \\ -\frac{1}{2} & 1 & \frac{1}{2} \end{pmatrix}\phi + \frac{v}{\Delta x}\begin{pmatrix} \frac{1}{2} & 0 & 0 \\ 1 & 1 & 0 \\ -\frac{1}{2} & -1 & 0 \end{pmatrix}\phi.$$

## 4.4 Other grids

The geometry of a problem often dictates the use of non-Cartesian coordinates, such as polars. There are enormous advantages in only using orthogonal coordinate systems, because differentials such as the Laplacian become rather complicated by the difference between co- and contravariant tensors.

Increased resolution of a small region with important activity can be obtained by using stretched grids, i.e. mappings $x(\xi)$ and/or $y(\eta)$. Note the independent stretches of the two coordinates, so that the coordinates remain orthogonal. It is much better to use central differencing on an equispaced grid in the mapped coordinates $\xi$ and $\eta$ with smooth mappings $x(\xi)$ and $y(\eta)$ than to use nonequispaced grids in $x$ and $y$. The disadvantage of using the two independent stretchings of the two coordinates is that increased resolution will occur wherever either coordinate is stretched while one may well wish to have increased resolution only in a corner where both are stretched.

Methods do exist for using localised increased resolution, e.g. Figure 4.2, but there are difficulties in successfully transferring information from the coarser grids to the finer and vice versa.

One general problem of stretched grids is that the stability of time-stepping is controlled by the smallest grid block, i.e. to avoid the diffusive numerical instability one requires

$$\Delta t < \tfrac{1}{4} Re \, \Delta x_{\min}^2,$$

and the advection instability

$$\Delta t < (\Delta x / U)_{\min}.$$

These restrictions become acute for polar coordinates

$$\Delta x_{\min} = r_{\min} \Delta \theta_{\min},$$

and $r_{\min} = \Delta r$ if the origin is in the computational domain. When working on the surface of a sphere, these very small spatial grid separations near the two poles can be avoided by patching together six equal spherical-square grids which have all grid blocks of similar size.

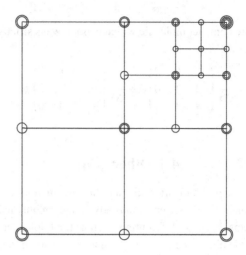

Figure 4.2 A series of grids with increased resolution in a corner.

In infinite domains, one sometimes needs to bring infinity nearer. This can be achieved with stretches such as

$$x = e^{\xi} \quad \text{or} \quad x = \frac{\xi}{1 - \xi}.$$

## 4.5  Conservative schemes

In §3.7 the algorithm gave slightly different values for the force on the top and the force on the sidewalls and bottom of the driven cavity. It is possible to devise an algorithm which conserves exactly the global momentum, so that the net force on all the walls vanishes. Conservative schemes are based on two ideas a conservative formulation of governing equations, and the application of those equations to Finite Volumes of Fluid.

The Navier–Stokes equations can be recast into the conservative form

$$\frac{\partial}{\partial t} (\rho \mathbf{u}) + \boldsymbol{\nabla} \cdot \mathbf{T} = 0$$

with total momentum flux

$$\mathbf{T} = \rho \mathbf{u} \mathbf{u} + p \mathbf{I} - 2\mu \mathbf{E}$$

combining Reynolds stresses, isotropic pressure and viscous stresses.

On the staggered grid of §3.6, the diagonal components of the stress are stored with the pressure at $i + \frac{1}{2} \, j + \frac{1}{2}$, and the off-diagonal component at $i \, j$,

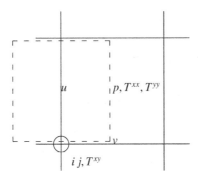

Figure 4.3 The staggered grid for $u_{i\,j+\frac{1}{2}}$, $v_{i+\frac{1}{2}\,j}$, $p_{i+\frac{1}{2}\,j+\frac{1}{2}}$ as before, and stresses $T^{xx}_{i+\frac{1}{2}\,j+\frac{1}{2}}$, $T^{yy}_{i+\frac{1}{2}\,j+\frac{1}{2}}$ and $T^{xy}_{i\,j}$; and in dash the Finite Volume centred on $i\,j+\frac{1}{2}$ for the $u$-momentum equation.

see Figure 4.3. For the $x$-component of momentum, the conservative form of the momentum equation is applied to the dashed volume marked in Figure 4.3, stretching from $x = (i - \frac{1}{2})\Delta x$ to $(i + \frac{1}{2})\Delta x$ and from $y = j\Delta x$ to $(j + 1)\Delta x$. The result is

$$\rho u^{n+1}_{i\,j+\frac{1}{2}} = \rho u^n_{i\,j+\frac{1}{2}} - \Delta t \left( \frac{T^{xx}_{i+\frac{1}{2}\,j+\frac{1}{2}} - T^{xx}_{i-\frac{1}{2}\,j+\frac{1}{2}}}{\Delta x} + \frac{T^{xy}_{i\,j+1} - T^{xy}_{i,j}}{\Delta x} \right).$$

When adding the changes in the momentum over the internal points, all the fluxes at internal boundaries between two adjacent finite volumes precisely cancel, leaving just fluxes at the outside. The sum of finite volumes for the $u$-momentum extends from the bottom to the top surface, while it falls one half-volume short at the two side walls. To achieve conservation of momentum for the entire volume, it is necessary to set boundary conditions, effectively for pressure on the wall, e.g. on $x = 0$

$$T^{xx}_{-\frac{1}{2}\,j+\frac{1}{2}} = T^{xx}_{\frac{1}{2}\,j+\frac{1}{2}} + T^{xy}_{0,j+1} - T^{xy}_{0\,j},$$

and similarly on the other walls.

The Reynolds stresses have to be evaluated by suitable averages over velocities held at other locations on the staggered grid. The viscous stresses can be evaluated without any averaging.

$$T^{xx}_{i+\frac{1}{2}\,j+\frac{1}{2}} = \rho\left(\frac{u_{i+1\,j+\frac{1}{2}} + u_{i\,j+\frac{1}{2}}}{2}\right)^2 + p_{i+\frac{1}{2}\,j+\frac{1}{2}} - 2\mu\frac{u_{i+1\,j+\frac{1}{2}} - u_{i\,j+\frac{1}{2}}}{\Delta x},$$

$$T^{xy}_{ij} = \rho\left(\frac{u_{i\,j+\frac{1}{2}} + u_{i\,j-\frac{1}{2}}}{2}\right)\left(\frac{v_{i+\frac{1}{2}\,j} + v_{i-\frac{1}{2}\,j}}{2}\right)$$
$$- \mu\left(\frac{u_{i\,j+\frac{1}{2}} - u_{i\,j-\frac{1}{2}}}{\Delta x} + \frac{v_{i+\frac{1}{2}\,j} - v_{i-\frac{1}{2}\,j}}{\Delta x}\right),$$

$$T^{yy}_{i+\frac{1}{2}\,j+\frac{1}{2}} = \rho\left(\frac{v_{i+\frac{1}{2}\,j+1} + v_{i+\frac{1}{2}\,j}}{2}\right)^2 + p_{i+\frac{1}{2}\,j+\frac{1}{2}} - 2\mu\frac{v_{i+\frac{1}{2}\,j+1} - v_{i+\frac{1}{2}\,j}}{\Delta x}.$$

The idea of a conservative form should be used in curvilinear coordinates, so that the form

$$\nabla^2\phi = \frac{1}{r^2}\frac{\partial}{\partial r}\left(r^2\frac{\partial\phi}{\partial r}\right) + \frac{1}{r^2}\frac{\partial^2\phi}{\partial\theta^2}$$

is better numerically than the theoretical equivalent

$$\frac{\partial^2\phi}{\partial r^2} + \frac{2}{r}\frac{\partial\phi}{\partial r} + \frac{1}{r^2}\frac{\partial^2\phi}{\partial\theta^2},$$

as can be seen by the discretisation

$$\frac{1}{r^2}\frac{\partial}{\partial r}\left(r^2\frac{\partial\phi}{\partial r}\right) \approx \frac{\left(r^2\frac{\partial\phi}{\partial r}\right)_{i+\frac{1}{2}} - \left(r^2\frac{\partial\phi}{\partial r}\right)_{i-\frac{1}{2}}}{r_i^2\Delta r} \quad \text{with} \quad \left(r^2\frac{\partial\phi}{\partial r}\right)_{i+\frac{1}{2}} \approx r_{i+\frac{1}{2}}^2\frac{\phi_{i+1} - \phi_i}{\Delta r}.$$

For two-phase flows, e.g. water and air, the so-called Volume of Fluid Method or One-Fluid Method is a conservative scheme treating the two phases as a single fluid with a density $\rho$ and viscosity $\mu$ which varies according to whether the part of a volume is one phase or the other.

The advection term in the Navier–Stokes equation has several alternative forms for conserving different properties.

$$\mathbf{u}\cdot\nabla\mathbf{u} = \nabla\cdot\mathbf{u}\,\mathbf{u} \qquad \text{conserves momentum}$$
$$= \nabla\tfrac{1}{2}u^2 - \mathbf{u}\wedge\omega \qquad \text{rotational form}$$
$$= \tfrac{1}{2}\mathbf{u}\cdot\nabla\mathbf{u} + \tfrac{1}{2}\nabla\cdot\mathbf{u}\,\mathbf{u} \qquad \text{conserves energy.}$$

The last form is known as the 'skew-symmetric differential form'. If the scalar product with $\mathbf{u}$ is taken with the last form, one obtains

$$\left(u_i u_j\left(u_i^{j+1} - u_i^{j-1}\right) + u_i\left(u_j^{j+1}u_i^{j+1} - u_j^{j-1}u_i^{j-1}\right)\right)/2\Delta x,$$

where the subscripts refer to components and the superscripts to location. On summing across the grid, the first and fourth terms cancel at internal points and the second and third terms cancel. Hence this form just rearranges energy without changing the total quantity. Conserving energy restricts the solution from blowing up everywhere (it could still blow up in a region of decreasing size), whereas fixed finite momentum does not restrict the growth of large cancelling positive and negative momentum.

# 5

---

# Finite elements

Finite elements are good for engineering problems with complex geometries. One has 'only' to triangulate the domain and then it is routine to apply the governing equations using the method. Finite elements were developed first for elliptic equations and are very good for them. They are reasonably good for parabolic equations but usually poor for hyperbolic equations. Finite elements naturally yield conservative schemes, and are often more accurate than they deserve to be.

The disadvantages are that it is difficult to generate the grid, difficult to write programs on unstructured grids, difficult to solve a Poisson problem efficiently on an unstructured grid and difficult to present the answers on an unstructured grid. These four difficulties are considerable. Hence it is best to leave the writing of codes to professionals, who have produced some good packages. The purpose of this chapter is therefore to explain the issues that those codes have tackled rather than give details on how to write code.

## 5.1 The two ideas

The finite elements approach is based on two ideas. The first is to represent the unknown functions everywhere within each element, i.e. everywhere within the domain. This should be contrasted with the finite difference approach of knowing the unknowns only at grid points. The following two sections are devoted to different possible representations of the unknown functions.

The second idea is a so-called *weak formulation* of the governing equations. This often comes down to a variational statement of the governing problem. This idea is explored in later sections.

## 5.2 Representations in one dimension

We divide the one-dimensional domain $a < x < b$ into $n$ segments with $a = x_0 < x_1 < \cdots < x_n = b$ to produce $n$ elements $x_{i-1} < x < x_i$ for $i = 1, 2, \ldots, n$. Note the segments need not have the same length.

### 5.2.1 Constant elements

At the crudest level, we can represent the unknown function $f(x)$ as constant within each element

$$f(x) = f_i \quad \text{in} \quad x_{i-1} \leq x < x_i,$$

as in Figure 5.1a. Note that the $f_i$ should not be thought of as values of the function at grid point $x_i$ as in finite differences but rather as just a parameter associated with the representation of the function over the element.

### 5.2.2 Linear elements

Better than piecewise constant is a piecewise linear representation of the unknown function $f(x)$

$$f(x) = f_{i-1}\frac{x_i - x}{x_i - x_{i-1}} + f_i\frac{x - x_{i-1}}{x_i - x_{i-1}} \quad \text{in} \quad x_{i-1} \leq x < x_i,$$

as in Figure 5.1b. Again the $f_i$ are parameters of the representation rather than spot values. If the same value of the parameter is used in the adjacent elements, the resulting representation would be continuous over the whole domain.

### 5.2.3 Quadratic elements

The next step in sophistication is obviously to represent the unknown function by a quadratic function over each element. Because the expressions now start to become complicated, it is a first worth mapping each element to a unit interval $0 \leq \xi \leq 1$ with

$$x(\xi) = x_{i-1} + (x_i - x_{i-1})\xi.$$

Then the representation by a quadratic is

$$f(x) = f_{i-1}(1 - \xi)(1 - 2\xi) + f_{i-\frac{1}{2}}4\xi(1 - \xi) + f_i\xi(2\xi - 1) \quad \text{in} \quad x_{i-1} \leq x < x_i.$$

While the representation is smooth within each element, there is still a discontinuity in the derivative of the representation across the boundary from one element to the next.

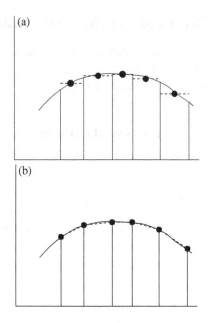

Figure 5.1 Finite elements in one dimension with (a) constant elements and (b) continuous linear elements.

### 5.2.4 Cubic elements

There are two choices of representation by a cubic over each element. The first

$$f(x) = f_{i-1}(1 - 3\xi)\left(1 - \frac{3}{2}\xi\right)(1 - \xi) + f_{i-\frac{2}{3}}\theta_2(\xi) + f_{i-\frac{1}{3}}\theta_1(\xi) + f_i\theta_0(\xi),$$

with obvious expressions for the various $\theta(\xi)$. While this cubic can better follow the unknown function within the element, it still has a discontinuity in the derivative of the representation across the boundary from one element to the next.

A second representation

$$f(x) = f_{i-1}(1 - \xi)^2(1 + 2\xi) + f'_{i-1}(1 - \xi)^2\xi(x_i - x_{i-1})$$
$$+ f_i\xi^2(3 - 2\xi) + f'_i\xi^2(1 - \xi)(x_i - x_{i-1}),$$

in terms of the four parameters $f_{i-1}, f'_{i-1}, f_i$ and $f'_i$, will be both continuous and have a continuous derivative across the boundary from one element to another so long as the parameters $f_i$ and $f'_i$ are the same for the two adjacent elements. The second derivative $f''(x)$ of the representation would be discontinuous at the boundary between elements. (See §11.1.1 about splines as a way of using cubic approximations with continuous second derivatives.)

### 5.2.5 Basis functions

In all the cases above, we can write the representation over the *whole domain* as the sum of products of simple basis functions $\phi_i(x)$ multiplied by unknown parameters or 'amplitudes' $f_i$

$$f(x) = \sum f_i \phi_i(x) \quad \text{in} \quad a < x < b.$$

The basis functions are zero through most of the domain, being nonzero in one or two elements only. Note that the parameters now labelled as just $f_i$ include the $f_i'$ in the second cubic representation.

For the constant elements, the basis functions are

$$\phi_i(x) = \begin{cases} 1 & \text{in} \quad x_{i-1} \leq x < x_i, \\ 0 & \text{otherwise.} \end{cases}$$

For the linear elements, the basis functions are

$$\phi_i(x) = \begin{cases} \dfrac{x - x_{i-1}}{x_i - x_{i-1}} & \text{in} \quad x_{i-1} \leq x \leq x_i, \\ \dfrac{x_{i+1} - x}{x_{i+1} - x_i} & \text{in} \quad x_i \leq x \leq x_{i+1}, \\ 0 & \text{otherwise,} \end{cases}$$

with obvious modifications for the end elements. See Figure 5.2.

For the second cubic elements with continuous derivative across boundaries, there are two types of basis functions, one $\phi_i(x)$ associated with the parameter $f_i$ and the other $\tilde{\phi}_i(x)$ associated with $f_i'$.

$$\phi_i(x) = \begin{cases} (x_{i+1} - x)^2(x_{i+1} + 2x - 3x_i)/(x_{i+1} - x_i)^3 & \text{in} \quad x_i \leq x < x_{i+1}, \\ (x - x_{i-1})^2(3x_i - 2x - x_{i-1})/(x_i - x_{i-1})^3 & \text{in} \quad x_{i-1} \leq x < x_i, \\ 0 & \text{otherwise,} \end{cases}$$

$$\tilde{\phi}_i(x) = \begin{cases} (x - x_i)(x_{i+1} - x)^2/(x_{i+1} - x_i)^2 & \text{in} \quad x_i \leq x < x_{i+1}, \\ (x - x_i)(x - x_{i-1})^2/(x_i - x_{i-1})^2 & \text{in} \quad x_{i-1} \leq x < x_i, \\ 0 & \text{otherwise.} \end{cases}$$

See Figure 5.3.

## 5.3 Representations in two dimensions

We start with representations over triangular elements with straight sides.

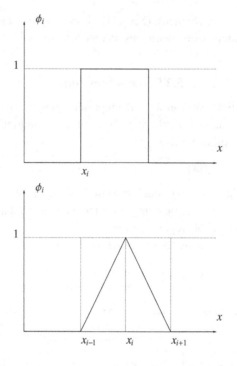

Figure 5.2  Basis functions for constant elements and linear elements.

### 5.3.1  Constant elements

The simplest representation of an unknown function is by a different constant in each element

$$f(x) = f_i \quad \text{in each triangle } i.$$

### 5.3.2  Linear elements

For linear elements, we need the linear interpolation functions $\ell_{12}(\mathbf{x})$, $\ell_{23}(\mathbf{x})$ and $\ell_{31}(\mathbf{x})$, where each function vanishes at two vertices and is unity on the other vertex, i.e. $\ell_{12}(\mathbf{x}_1) = 0 = \ell_{12}(\mathbf{x}_2)$ and $\ell_{12}(\mathbf{x}_3) = 1$, where $\mathbf{x}_1, \mathbf{x}_2$ and $\mathbf{x}_3$ are the three vertices. Hence

$$\ell_{12}(x, y) = \frac{(x - x_1)(y_2 - y_1) - (x_2 - x_1)(y - y_1)}{(x_3 - x_1)(y_2 - y_1) - (x_2 - x_1)(y_3 - y_1)}.$$

With these three linear interpolation functions, an unknown function can be represented linearly within one element as

$$f(\mathbf{x}) = f_1 \ell_{23}(\mathbf{x}) + f_2 \ell_{31}(\mathbf{x}) + f_3 \ell_{12}(\mathbf{x}).$$

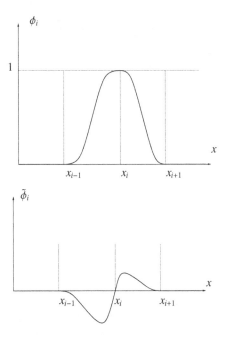

Figure 5.3 Basis for cubic elements with continuous first derivative.

By making the vertex values $f_i$ the same for elements sharing that vertex, the representation becomes continuous over the whole domain.

### 5.3.3 Quadratic elements

For a quadratic representation of the unknown functions, we can use products of the above linear interpolations functions. Thus

$$f(\mathbf{x}) = f_1 \ell_{23}(\mathbf{x})(2\ell_{23}(\mathbf{x}) - 1) + f_2 \ell_{31}(\mathbf{x})(2\ell_{31}(\mathbf{x}) - 1) + f_3 \ell_{12}(\mathbf{x})(2\ell_{12}(\mathbf{x}) - 1)$$
$$+ f_{23} 4\ell_{12}(\mathbf{x})\ell_{31}(\mathbf{x}) + f_{31} 4\ell_{23}(\mathbf{x})\ell_{12}(\mathbf{x}) + f_{12} 4\ell_{31}(\mathbf{x})\ell_{23}(\mathbf{x}).$$

This representation takes the value $f_1$ at the vertex $\mathbf{x}_1$ and the value $f_{23}$ at the midpoint along the edge from $\mathbf{x}_2$ to $\mathbf{x}_3$, see Figure 5.4. By making the vertex values and midpoint values the same for vertices and midpoints shared by two elements, the representation becomes continuous and a continuous tangential derivative along the edges. The normal derivative to an edge is however discontinuous, and the derivative is quite unpleasant at a vertex.

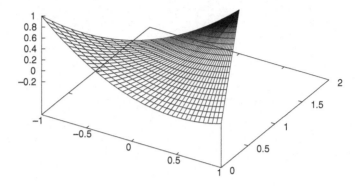

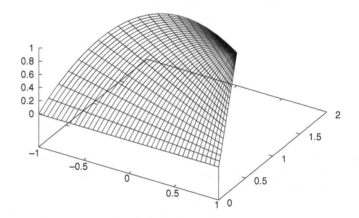

Figure 5.4 The quadratic elements in two dimensions, $\ell_{23}(\mathbf{x})(2\ell_{23}(\mathbf{x}) - 1)$ and $4\ell_{12}(\mathbf{x})\ell_{23}(\mathbf{x})$.

### 5.3.4 Cubic elements

A general cubic in two dimensions has 10 degrees of freedom. With this flexibility, it is possible to construct a cubic representation which takes given values and given (two-dimensional) derivatives at each vertex of a triangle, leaving one degree of freedom unused. This latter could be taken as something like the value at the centre of the triangle (and vanishing with vanishing derivatives at all the vertices), which is called a 'bubble' function.

### 5.3.5 Basis functions

Adopting one of the above representations, or a further generalisation, over each element, we can write the representation over the whole domain in terms of amplitudes $f_i$ and basis functions $\phi_i(\mathbf{x})$

$$f(\mathbf{x}) = \sum_i f_i \phi_i(\mathbf{x}).$$

Note again that the amplitudes will normally not be a value of the unknown function at any special point. Also the basis functions will be nonzero only in a small number of elements. For example for linear elements, the basis function associated with one vertex will be nonzero only in the triangles which share that particular vertex. As the linear basis function will vanish along the edges opposite the particular vertex, it will take the form of a several-sided pyramid. The local nature of the basis functions helps create sparse coupling matrices when the finite elements are applied to nonlinear partial differential equations.

### 5.3.6 Rectangles

For complex geometries, the domain is most easily divided into triangular elements. In simpler geometries, rectangular elements are sometimes used.

The simplest representation of an unknown functions is by different constants in each element.

It is not possible to make a linear interpolation between the values at the four corners of a rectangle. Instead a so-called *bilinear* form is used which is separately linear in the two coordinates. Suppose the $xy$-rectangle has been mapped to the unit square $0 \le \xi \le 1$, $0 \le \eta \le 1$. Then the bilinear representation would be

$$f(\mathbf{x}) = f_1 \xi \eta + f_2 (1 - \xi) \eta + f_3 \xi (1 - \eta) + f_4 (1 - \xi)(1 - \eta).$$

Note the only quadratic term is the product $\xi \eta$. The parameters $f_i$ are the values of the representation in one of the four corners. By using the same values for all elements sharing a corner, one constructs a representation which is continuous over the whole domain.

One can similarly construct a biquadratic representation which is the sum of nine terms, each the product of separate quadratics in the two coordinates. One possibility is to make each of the terms vanish at all but one of the corners or midpoints. Again using the same values of the parameters for elements sharing the same corners of midpoints, one constructs a representation which is continuous over the whole domain, and which has a continuous tangential derivative along each edge, although the normal derivatives are not continuous and the derivative is not nice at a corner.

The one- and two-dimensional representations can be generalised in an obvious way to three dimensions. One can also used curved elements by first mapping them onto straight-sided elements.

## 5.4 Variational statement of the Poisson problem

The Poisson problem arises in many branches of physics and is an illustration of more general self-adjoint elliptic problems:

$$\nabla^2 f = \rho \quad \text{in volume } V$$

with boundary condition, say $\quad f = g \quad$ on surface $S$,

with $\rho(\mathbf{x})$ and $g(\mathbf{x})$ given. For this problem, we have a Rayleigh–Ritz variational formulation: out of all those functions $f(\mathbf{x})$ which satisfy the boundary conditions, the one which minimises

$$I(f) = \int_V \left( \frac{1}{2} |\nabla f|^2 + \rho f \right) dV$$

also satisfies the Poisson problem.

Into this variational statement we substitute our finite element representation

$$f(\mathbf{x}) = \sum f_i \phi_i(\mathbf{x}).$$

Then

$$I(f) = \frac{1}{2} \sum_{ij} f_i f_j \int \nabla \phi_i \cdot \nabla \phi_j + \sum_i f_i \int \rho \phi_i.$$

We define the *global stiffness matrix*

$$K_{ij} = \int \nabla \phi_i \cdot \nabla \phi_j,$$

and *forcing*

$$r_i = \int \rho \phi_i.$$

Then minimising $I(f)$ over the amplitudes $f_i$, we obtain an equation for the unknown amplitudes

$$K_{ij} f_j + r_i = 0.$$

Note with the $f_i$ determined by this equation, the representation $f(\mathbf{x}) = \sum f_i \phi_i(\mathbf{x})$ satisfies

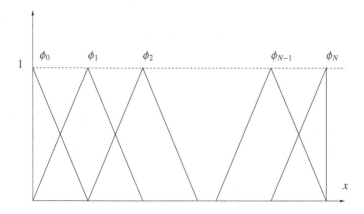

Figure 5.5 Linear elements in one dimension.

$$- \int \nabla f \cdot \nabla \phi_j = \int \rho \phi_j \quad \text{for all } j,$$

i.e. the governing equation is satisfied in each of the finite directions accessible to the finite element representation. This is called a *week formulation* of the partial differential equation. It can be used, and indeed will shortly be used, with representations $f(\mathbf{x})$ which are continuous and have a derivative $\nabla f$, so that the weak formulation can be considered, but which have no nice $\nabla^2 f$ (along the edges), so that the original partial differential equation has no simple meaning.

## 5.5 Details in one dimension

To see the finite element method working in practice, consider the Poisson problem in one-dimensional for the unknown $f(x)$ in

$$\frac{d^2 f}{dx^2} = \rho \quad \text{in } a < x < b$$

with $f(a) = A$ and $f(b) = B$,

where $\rho(x), A$ and $B$ are given.

We divide the interval $[a, b]$ into $N$ equal segments of length $h = (b - a)/N$. We use linear finite elements with basis functions as in Figure 5.5. Hence the unknown $f(x)$ is represented

$$f(x) = A\phi_0(x) + B\phi_N(x) + \sum_{i=1}^{N-1} f_i \phi_i(x)$$

in which the boundary conditions have been built in. At the interior points, $i, j = 1, 2, \ldots, N - 1$, the global stiffness matrix takes the values.

$$K_{ij} = \int \nabla \phi_i \cdot \nabla \phi_j = \begin{cases} 2/h & \text{if } i = j, \\ -1/h & \text{if } i = j \pm 1, \\ 0 & \text{otherwise.} \end{cases}$$

These results follow from the gradient being either zero or $\pm 1/h$, with agreement over two segments of length $h$ when $i = j$, with opposite signs over the one common segment when $i = j \pm 1$ and with no overlap of the nonzero segments otherwise. If we take the given $\rho(x)$ to be piecewise constant, then the forcing $r_i$ are given by

$$r_i = \int \rho(x)\phi_i = h\rho_i.$$

The equation governing the unknown amplitudes $f_i$ then becomes

$$\frac{1}{h}(-f_{i-1} + 2f_i - f_{i+1}) + h\rho_i = 0 \quad \text{for } i = 1, 2, \ldots, N - 1,$$

which is the same equation for the point values in the finite difference approach.

Note if the interval had been divided unequally into segments of length $h_{\frac{1}{2}}, h_{\frac{3}{2}}, \ldots, h_{N-\frac{1}{2}}$, then the finite element equation for the unknown amplitudes $f_i$ would have been

$$\frac{1}{h_{i-\frac{1}{2}}}(-f_{i-1} + f_i) + \frac{1}{h_{i+\frac{1}{2}}}(f_i - f_{i+1}) + \frac{h_{i-\frac{1}{2}} + h_{i+\frac{1}{2}}}{2}\rho_i = 0.$$

This form shows the finite element approach naturally produces a conservative scheme.

Note if the forcing integrals $r_i$ had been evaluated more accurately, we would have obtained

$$r_i = \int \rho(x)\phi_i(x) = \rho_i + \frac{h^3}{12}\rho_i'' + O(h^5).$$

The governing equation for these finite elements would then give an error of $O(h^4)$ in the representation on the boundaries of the elements, although the error at interior points would be $O(h^2)$.

## 5.6 Details in two dimensions

A more typical example of the finite element method working is the solution of the Poisson problem in two dimensions. We use linear basis functions on triangular elements, see Figure 5.6. Consider the basis function $\phi_1(\mathbf{x})$ which vanishes on the 23-side and is unity at the 1-vertex. The magnitude of its gradient is one upon the altitude $h_1$ from the 23-side to the 1-vertex, i.e.

$$|\nabla\phi_1| = \frac{1}{h_1}.$$

Hence the contribution of this triangle to the diagonal element of the global stiffness matrix is

$$K_{11} = \int \nabla\phi_1 \cdot \nabla\phi_1 = \frac{A}{h_1^2},$$

where $A$ is the area of the triangle. Now the angle between $\nabla\phi_1$ and $\nabla\phi_2$ is $\pi - \theta_3$, where $\theta_3$ is the included angle at the 3-vertex. Hence the off-diagonal element of the stiffness matrix has a contribution from this triangle of

$$K_{12} = \int \nabla\phi_1 \cdot \nabla\phi_2 = -\frac{\cos\theta_3 A}{h_1 h_2}.$$

Now by elementary geometry, the altitudes can be related to the lengths of the sides by

$$h_1 = \ell_2 \sin\theta_3 \quad \text{and} \quad h_2 = \ell_1 \sin\theta_3.$$

While the area can be expressed

$$A = \tfrac{1}{2}\ell_1 \ell_2 \sin\theta_3.$$

Hence the off-diagonal contribution simplifies to

$$K_{12} = -\frac{\cos\theta_3 A}{h_1 h_2} = -\tfrac{1}{2}\cot\theta_3.$$

Now the length of one side can be split into two parts

$$\ell_1 = h_1 \cot\theta_3 + h_1 \cot\theta_2.$$

Hence the diagonal contribution simplifies to

$$K_{11} = \frac{A}{h_1^2} = \tfrac{1}{2}\left(\cot\theta_3 + \cot\theta_2\right).$$

Note that the contributions from the triangle to the stiffness matrix sum to zero,

$$K_{11} + K_{12} + K_{13} = 0.$$

This follows from the sum of the basis functions being a linear function which is unity at each vertex, and so

$$\phi_1(\mathbf{x}) + \phi_2(\mathbf{x}) + \phi_3(\mathbf{x}) \equiv 1.$$

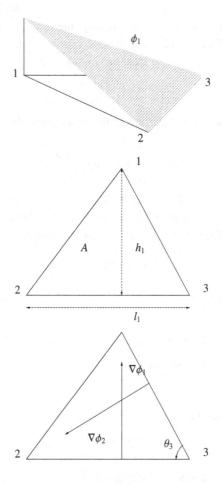

Figure 5.6 Linear element in two dimensions.

Being constant, the sum has zero gradient, i.e.

$$\nabla\phi_1 \cdot (\nabla\phi_1 + \nabla\phi_2 + \nabla\phi_3) \equiv 0.$$

We now assemble the contributions from the different triangles that meet at one vertex. For simplicity, we consider a square Cartesian grid with the squares divided by a diagonal to form triangular elements, see Figure 5.7.

For the 123-triangle, the contribution to $K_{13}$ involves the other (not 1 or 3) angle, $\theta_2 = \frac{\pi}{2}$, whose cotangent vanishes, so the contribution to $K_{13} = 0$. Similarly the contribution to $K_{12}$ involves $\theta_3 = \frac{\pi}{4}$, whose cotangent is unity, so $K_{12} = -\frac{1}{2}$. The contribution to the diagonal element is the negative sum of the off-diagonals, so $K_{11} = \frac{1}{2}$. Turning to the 172-triangle, we have contributions to

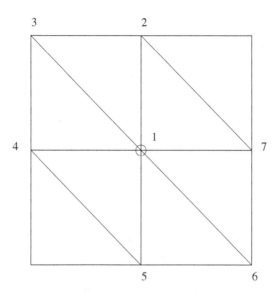

Figure 5.7 Assembled linear element in two dimensions.

the two off-diagonal elements equal, $K_{17} = K_{12} = -\frac{1}{2} \cot \frac{\pi}{4} = -\frac{1}{2}$. And by the diagonal element being the negative sum of the off-diagonal, the contribution to $K_{11} = 1$. Now all the triangles which meet at vertex 1 are similar to one or other of these triangles, so we can add up the contributions to find

$$K_{11} = 4, \quad K_{12} = K_{14} = K_{15} = K_{17} = -1, \quad K_{13} = K_{16} = 0.$$

We must now evaluate the forcing terms from $\rho$,

$$r_i = \int \rho \phi_i.$$

At the lowest approximation we take $\rho(\mathbf{x})$ to be constant $\rho_i$. Then integrating the linearly varying $\phi(\mathbf{x})$ over just one triangle, we have a contribution from the triangle of $\frac{1}{3} A \rho_i$. Adding up the contributions from the six triangles at vertex 1, and using $A = \frac{1}{2} h^2$ where $h$ is the size of the squares, we have

$$r_1 = h^2 \rho_1.$$

Hence the Poisson problem in finite elements

$$K_{ij} f_j + r_i = 0$$

becomes on our special triangular grid

$$\begin{pmatrix} 0 & -1 & 0 \\ -1 & 4 & -1 \\ 0 & -1 & 0 \end{pmatrix} f + h^2 \rho_i = 0,$$

which is identical to the finite difference equation when the amplitudes $f_i$ are the grid values.

On more general, unstructured grids there is a programming challenge of maintaining useful lists. One needs a list of points $P$, which have coordinates $(x_P, y_P)$ and an indicator of whether the point is interior or on the boundary. One needs a list of triangles $T$, which have vertices $P_{1T}, P_{2T}, P_{3T}$. This list can be searched to find in which triangle an arbitrary point lies. One needs the inverse list, i.e. for each vertex $P$ which $M$ triangles $T_{1P}, T_{2P}, \ldots, T_{MP}$ contain the vertex. An alternative to this list is a list of edges $E$ joining points $P_{1E}$ and $P_{2E}$. From either of these alternative lists, one can assemble the sparse stiffness matrix. There are other possible lists, whose usefulness depends on the particular problem being tackled, where the overheads of setting up and maintaining the lists has to be balanced by savings in CPU times and memory consumed.

## 5.7 Galerkin formulation

Many branches of physics have a variational formulation, in which the governing partial differential equations are the Euler–Lagrange equations of the minimisation of some expression for an Action. But some physics does not have a variational statement, e.g. the Navier–Stokes equations. In such cases, we use a so-called *weak formulation* of the problem, sometimes also called a *Galerkin formulation*.

Consider the governing equation written in the symbolic form

$$A(u) = f,$$

where $A$ is a nonlinear partial differential equation for the unknown $u(\mathbf{x}, t)$. We use a finite element representation

$$u(\mathbf{x}, t) = \sum_{i}^{N} u_i(t) \phi_i(\mathbf{x}),$$

with a finite number $N$ of amplitudes $u_i(t)$ and localised basis functions $\phi_i(\mathbf{x})$. Thus we can view our approximation to the exact solution to be a member of a finite vector space spanned by the basis functions. On this vector space, we define an inner product

$$\langle a, b \rangle = \int a(\mathbf{x})b(\mathbf{x})\, dV.$$

The Galerkin approximation is to require the residual in satisfying the equation, $A(u) - f$, to have no component in our solution space, i.e. to be orthogonal to all the basis functions

$$\langle A(u) - f, \phi_j \rangle = 0 \quad \text{all } j.$$

If the operator $A$ contains derivatives of order higher than the first, then one can integrate by parts to transfer half the derivatives on $u$ to operate instead on the $\phi_j$. For example in §5.4, the second-order derivative in the original Poisson equation for $f$ was reduced to first order only on the $\phi$ in the global stiffness matrix $K$. Indeed the basis functions used in the two previous sections were piecewise linear, so their first derivative was square-integrable in $K$ while their second derivatives existed only in the form of delta functions along the edges. This reduction in the differentiability requirements of the basis functions is the origin of the word *weak*: the finite element representation is required to satisfy *weaker* conditions than those necessary to satisfy the original governing equation. For those interested in the rigorous underpinning of the subject, one would need to prove (i) that a solution to the weak formulation of the Finite Element problem exists for any $N$, (ii) that the solutions converges to a limit as $N \to \infty$, and (iii) that the limit solution is nice and smooth, and finally (iv) that the limit solution satisfies the original governing equation.

Nearly always the basis functions can represent a constant function. With an appropriate normalisation, we would have

$$\sum \phi_j(\mathbf{x}) \equiv 1,$$

perhaps not including all the basis functions. Now the weak formulation is

$$\langle A(u) - f, \phi_j \rangle = 0 \quad \text{all } j.$$

Summing over the $j$, we therefore have

$$\langle A(u) - f, 1 \rangle = 0,$$

i.e.

$$\int A(u) = \int f.$$

Thus the finite element approach automatically satisfies a global conservation property.

## 5.8 Diffusion equation

### 5.8.1 Weak formulation

We consider the application of the weak formulation to the diffusion equation as a step towards the more complicated Navier–Stokes equation

$$u_t = \nabla^2 u.$$

The weak formulation is

$$\langle u_t - \nabla^2 u, \phi_j \rangle = 0 \quad \text{all } j.$$

Integrating by parts,

$$\langle u_t, \phi_j \rangle = -\langle \nabla u, \nabla \phi_j \rangle.$$

Substituting in the finite element representation,

$$u(\mathbf{x}, t) = \sum_i u_i(t)\phi_i(\mathbf{x}),$$

we obtain

$$\sum_i \dot{u}_i(t)\langle \phi_i, \phi_j \rangle = -\sum_i u_i(t)\langle \nabla \phi_i, \nabla \phi_j \rangle,$$

i.e.

$$M_{ij}\dot{u}_j = -K_{ij}u_j,$$

with new

$$\text{'Mass' matrix} \quad M_{ij} = \langle \phi_i, \phi_j \rangle,$$

and the earlier

$$\text{'Stiffness' matrix} \quad K_{ij} = \langle \nabla \phi_i, \nabla \phi_j \rangle.$$

### 5.8.2 In one dimension

Using the linear elements on equal intervals $h$, we have

$$M_{ij} = \begin{cases} \frac{2}{3}h & i = j, \\ \frac{1}{6}h & i = j \pm 1, \\ 0 & \text{otherwise,} \end{cases} \quad \text{and} \quad K_{ij} = \begin{cases} 2/h & i = j, \\ -1/h & i = j \pm 1, \\ 0 & \text{otherwise.} \end{cases}$$

Hence at an interior point we obtain the equation governing the evolution of the amplitudes

$$h\left(\tfrac{1}{6}\dot{u}_{i-1} + \tfrac{2}{3}\dot{u}_i + \tfrac{1}{6}\dot{u}_{i+1}\right) = \frac{1}{h}\left(u_{i-1} - 2u_i + u_{i+1}\right).$$

Note this leaves a linear algebra problem to find the $\dot{u}_i$. The tridiagonal matrix $M$ can however be rapidly inverted.

The equation above is called a 'semidiscretised' form, a form in which the spatial variation has been discretised but which leaves open the method of time integration. One could use a simple forward time-stepping

$$u_i^{n+1} = u_i^n + \Delta t \dot{u}_i^n,$$

or one of a number of more refined schemes. However it is certainly very unwise to discretise the time-stepping using finite elements in time.

### 5.8.3 In two dimensions

We again use the linear basis functions on triangular elements, see Figure 5.6. The contributions from a triangle to the elements of the mass matrix are

$$M_{ij} = \begin{cases} \frac{1}{12}h^2 & i = j, \\ \frac{1}{24}h^2 & i \neq j, \end{cases}$$

and the same stiffness matrix as in §5.6. Hence assembling the contributions from the different triangles in Figure 5.7, we obtain

$$\tfrac{1}{2}h^2 \left( \dot{u}_1 + \tfrac{1}{6}(\dot{u}_2 + \dot{u}_3 + \dot{u}_4 + \dot{u}_5 + \dot{u}_6 + \dot{u}_7) \right) = u_2 + u_4 + u_5 + u_7 - 4u_1.$$

Again there is a linear algebra problem to find $\dot{u}_i$.

## 5.9 Navier–Stokes equation

### 5.9.1 Weak formulation

We use a finite element representation for the velocity $\mathbf{u}$ and pressure $p$,

$$\mathbf{u}(\mathbf{x}, t) = \sum_i \mathbf{u}_i(t)\phi_i(\mathbf{x}),$$

$$p(\mathbf{x}, t) = \sum_i p_i(t)\psi_i(\mathbf{x}),$$

with unknown amplitudes $\mathbf{u}_i$ and $p_i$ and basis functions $\phi_i$ and $\psi_i$. Note that one normally uses different basis functions for the velocity and pressure, for example the pressure with one less derivative than the velocity. The weak formulation of the Navier–Stokes equation is then

$$\left\langle \rho \left( \frac{\partial \mathbf{u}}{\partial t} + \mathbf{u} \cdot \nabla \mathbf{u} \right) + \nabla p - \mu \nabla^2 \mathbf{u}, \phi_j \right\rangle = 0 \quad \text{all } \phi_j,$$

and the incompressibility constraint

$$\langle \nabla \cdot \mathbf{u}, \psi_j \rangle = 0 \quad \text{all } \psi_j.$$

Integrating the viscous term by parts, we obtain the system of equations governing the amplitudes

$$\rho \left( M_{ij}\dot{\mathbf{u}}_j + Q_{ijk}\mathbf{u}_j\mathbf{u}_k \right) = -B_{ji}p_j - \mu K_{ij}\mathbf{u}_j,$$

and

$$-B_{ij}\mathbf{u}_j = 0,$$

with the earlier mass $M$ and stiffness $K$ matrices and two new coupling matrices

$$Q_{ijk} = \langle \phi_i \nabla \phi_j, \phi_k \rangle \quad \text{and} \quad B_{ij} = \langle \nabla \psi_i, \phi_j \rangle = -\langle \psi_i, \nabla \phi_j \rangle.$$

### 5.9.2  Time integration

The time marching of the above semidiscretised form can be made using one's favourite finite difference (in time) method. e.g. the simplest explicit forward step

$$\mathbf{u}_i^{n+1} = \mathbf{u}_i^n + \Delta t \dot{\mathbf{u}}_i^n.$$

The incompressibility can be taken into account by a projection split step method, as in §3.4, i.e.

$$\mathbf{u}^* = \mathbf{u}_i^n + \Delta t \left( \dot{\mathbf{u}}_i^n \quad \text{without the } p \text{ term} \right),$$
$$\mathbf{u}^{n+1} = \mathbf{u}^* + \Delta t \left( \dot{\mathbf{u}}_i^n \quad \text{with just the } p \text{ term} \right),$$

with $p$ chosen so the incompressibility is satisfied at the end of the step

$$B\mathbf{u}^{n+1} = 0.$$

A second order $O(\Delta t^2)$ pressure update method is better than this pressure projection method, see §7.10.

### 5.9.3  Pressure problem – locking

Consider using triangular elements with velocity linear and pressure constant over each element. The pressure basis functions $\psi_j(\mathbf{x})$ will be unity in triangle $\Delta_j$ and zero in all the other triangles. The weak formulation of the incompressibility constraint,

$$\langle \nabla \cdot \mathbf{u}, \psi_j \rangle = 0 \quad \text{all } j,$$

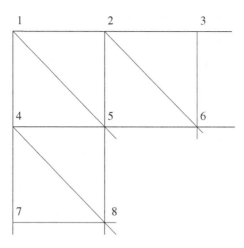

Figure 5.8 Pressure locking.

then gives

$$\oint_{\Delta_j} u_n = 0,$$

i.e. no net volume flux out of triangle $\Delta_j$.

We now examine the consequences of this condition of no net volume flux out of each triangle on the grid in Figure 5.8, of squares of side $h$ with diagonals added to form triangular elements. The no-slip boundary condition of both components of velocity vanishing is applied to the boundary vertices 1, 2, 3, 4 and 7. Let $(u_5, v_5)$ be the unknown amplitude of the velocity at vertex 5. Consider triangle 145. With the velocity varying linearly along the edge 45, the flux in across edge 45 is $\frac{1}{2}hv_5$. The flux out across edge 15 is however $\frac{1}{2}h(u_5+v_5)$. Hence the zero net mass flux constraint gives $u_5 = 0$. Now consider the triangle 125. The net flux into this triangle is $\frac{1}{2}hv_5$, whose vanishing gives $v_5 = 0$. Now that both components of velocity vanish at vertex 5, we can use the same argument to deduce that both components must vanish at vertices 6 and 8. We can continue this process right across the grid, deducing that the velocity must vanish everywhere. This deduction follows from the constraint of no net mass flux into each triangle and the no-slip boundary condition. The momentum equation has yet to be considered.

This paradoxical behaviour becomes easier to understand if we look at the whole domain rather than one triangle. For one triangle we seem to have one unknown pressure amplitude and six unknown velocity component amplitudes (two components at three vertices), i.e. many more degrees of freedom in the

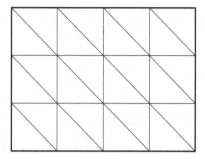

Figure 5.9 Locking due to many triangles and few interior points.

velocity than the pressure. Now consider the small grid in Figure 5.9 of four squares by three squares. There are 24 triangles, each with a pressure amplitude. There are however only 6 interior vertices with therefore only 12 unknown velocity component amplitudes. The problem thus comes from each vertex being shared by six triangles. To redress the imbalance, one can introduce so-called bubble functions for both velocity components. These basis functions vanish at every vertex and are unity at the centre of the triangle, and are hence cubic in nature. This introduces two extra unknown velocity component amplitudes to each triangle, which means there are now more degrees of freedom for the momentum equation than the incompressibility constraint. Typical practice is to use triangular elements with either linear pressure with linear velocity plus a cubic velocity bubble or slightly better linear pressure with quadratic velocity plus a cubic velocity bubble.

### 5.9.4 Pressure problem – spurious modes

As in §3.5 on finite differences, finite elements can have spurious pressure modes which do not contribute to the momentum equation. In fact, they are nearly unavoidable in finite elements, which does not have an equivalent to finite differences' staggered grid. Consider pressure linear over the triangles in Figure 5.9. If the pressure is ±1 with alternating signs at adjacent vertices along the horizontal and vertical, then around one vertex there is a pressure distribution as is in Figure 5.10. By symmetry this distribution of pressure has no pressure gradient at the vertex, and hence does not contribute to the momentum equation.

The problem of spurious pressure modes in finite elements reduces to the

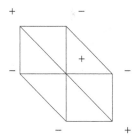

Figure 5.10 A spurious pressure mode.

coupling matrix $B$ having eigensolutions, i.e. nontrivial pressures $p_j$ such that

$$B_{ji}p_j = 0.$$

One would like to find basis functions for the velocity and pressure which produced a coupling matrix $B$ which had no eigensolutions. This requirement is also called the Babuška–Brezzi condition.

If a set of well-behaved basis functions cannot be used, an alternative method of suppressing the spurious pressure modes is to modify the physics with a 'pressure penalty', replacing the incompressibility equation by

$$\nabla \cdot \mathbf{u} = -\beta h^2 p$$

where $h$ is the size of the elements and $\beta$ is a constant, whose optimal value is thought to be $\beta = 0.025$. The weak formulation of the modified incompressibility condition is

$$B_{ij}\mathbf{u}_j - \beta h^2 p_i = 0.$$

This equation has the desirable feature of bringing in the pressure whose role is to help achieve incompressibility but which does not normally occur in the constraint.

## Further reading

*Finite elements for Navier–Stokes equations* by R. Glowinski and O. Pironneau in Annu. Rev. Fluid Mech. (1992) **24**, 167–204.

# 6

# Spectral methods

Spectral methods are for

- very simple geometry, e.g. Cartesian, and
- very smooth functions, e.g. $C^\infty$-functions with all derivatives continuous and so bounded.

Under these conditions spectral methods are remarkably accurate, with errors decreasing faster than any power of the number of modes $N$, typically decreasing like $e^{-kN}$. Most waves need only three modes to represent them with, say, a 1% accuracy, whereas second-order finite differences would need about 40 points and fourth-order 20 points. Differentiation is exact to the shortest wave, and the Poisson solver is trivial.

The shortcoming is that the spectral transformation is time consuming, and this is needed to treat otherwise very expensive nonlinear terms. Sometimes there are so-called Fast Transforms, which then make spectral methods competitive.

## 6.1 The two ideas

A spectral method approach is based on two ideas, just like finite elements. The first is a spectral representation of the unknown functions, and the second is a Galerkin approximation to solving the partial differential equation. The spectral representation takes the form

$$u(x,t) = \sum_n^N \hat{u}_n(t)\phi_n(x),$$

with a finite number of amplitudes $\hat{u}_n(t)$ and basis functions $\phi_n(x)$, such as trigonometric functions in a Fourier representation.

For a partial differential equation written in the symbolic form

$$A(u) = f,$$

the Galerkin approximation is to chose the solution $u(x, t)$ such that the residual in satisfying the governing equation is orthogonal to each of the finite number of basis functions, i.e.

$$\langle A(u) - f, \phi_n \rangle = 0 \quad \text{for} \quad n = 0, 1, \ldots, N.$$

In other words, the residue vanishes in all the directions that the set of basis functions can look at.

## 6.2 Global vs local

Consider the case of Fourier transforms over an infinite interval

$$u(x) = \int e^{ikx} \hat{u}(k) \, dk, \qquad \hat{u} = \frac{1}{2\pi} \int e^{-ikx} u(x) \, dx.$$

*Differentiation* is global in real space, in the sense that point values of the function throughout the interval have an influence on the derivative at one point, albeit a weak influence from remote points. On the other hand, the transform of the derivative

$$\widehat{\frac{du}{dx}}(k) = ik\hat{u}(k)$$

is a local operator involving only the transform of the function at that particular mode $k$.

Note that spectral differentiation is exact for the highest mode, $u(x) = \sin(\pi * x/\delta x)$, whereas central differences would give zero, $(f_{i+1} - f_{i-1})/2\delta x = 0$.

The *Poisson problem* is similarly global in real space

$$\frac{d^2 u}{dx^2} = \rho,$$

in which the answer at one point $u(x)$ depends on values of $\rho(x')$ at all points $x'$. On the other hand taking Fourier transforms, the problem is

$$-k^2 \hat{u} = \hat{\rho},$$

which is immediately invertible, with $\hat{u}(k)$ only involving $\hat{\rho}$ at the same single $k$, i.e. a local problem.

With *nonlinear terms* and terms with *varying coefficients*, the above useful property of a spectral approach falls apart. While the product

$$u(x)v(x)$$

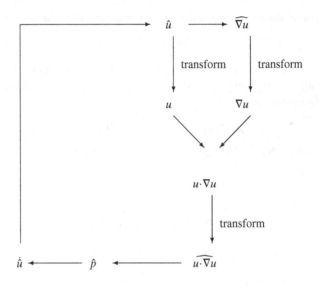

Figure 6.1 A pseudospectral approach with manipulations in real and in Fourier space.

is local in real space, it becomes global in Fourier space

$$\widehat{uv}(k) = \frac{1}{2\pi} \int_{l+m=k} \hat{u}(l)\hat{v}(m).$$

For numerical calculations, *local is cheap and global is expensive.*

The Navier–Stokes equation unfortunately involves a combination of local and global features whether tacked in real space or Fourier space.

A mix of manipulations in real and in Fourier space can therefore be beneficial. Such a mix is called a *pseudospectral method.* Thus one can rapidly evaluate the derivative $\widehat{\nabla u}$ in Fourier space, then transform this and $\hat{u}$ to real space $\nabla u(x)$ and $u(x)$, form the product $u \cdot \nabla u$ in real space, transform back to $\widehat{u \cdot \nabla u}$ in Fourier space, and then go on to solve (rapidly) the Poisson problem in order to time-step the transform $\hat{u}$, see Figure 6.1. Going with the use of a finite number of Fourier modes, one uses a finite number of so-called *collocation* points in real space. The choice of the optimal location of these points depends on the choice of the spectral basis function, but tends to be the zeros of the first basis function beyond the last used, $\phi_{N+1}$. An alternative pseudospectral strategy would be to time-step in real space, transforming $u(x)$ to Fourier space for the fast differentiation, transforming back to real space to form the product $u \cdot \nabla u$, transforming back to invert the Poisson problem and then transforming again to time-step in real space. This alternative involves too many transforms.

## 6.3 Choice of spectral basis functions

Spectral basis functions $\phi_n(x)$ should have the following properties:

- be complete, i.e. can represent all the functions of interest,
- be orthogonal, i.e.

$$\langle \phi_m, \phi_n \rangle = \int \phi_m(x) \phi_n(x) w(x) \, dx = N_n \delta_{mn},$$

  with some appropriate weight function $w$,
- be smooth, which affects
- have fast convergence, so that few terms are needed,
- have a Fast Transform if possible, and
- satisfy the boundary conditions if possible.

It is strongly recommended that for fully periodic problems on the interval $[0, 2\pi]$ one uses for the basis functions the Fourier series $e^{in\theta}$. Using the complex exponential form rather than sines and cosines avoids an awkward exceptional factor of $\frac{1}{2}$ in the zero mode. Fully periodic means that the function and all its derivatives at the end of the interval are equal to the values at the start of the interval

$$f^{(k)}(2\pi-) = f^{(k)}(0+) \quad \text{all } k.$$

Fully periodic functions occur naturally as the solutions of differential equation with periodic boundary conditions.

It is strongly recommended that for problems on the finite interval $[-1, 1]$ which are not fully periodic one uses Chebyshev polynomials $T_n(\cos\theta) = \cos n\theta$.

In some non-Cartesian geometries, there are suitable orthogonal functions with which to form a spectral basis, e.g. Bessel functions in cylindrical geometry and Legendre polynomials in spherical geometry. While these functions are complete, orthogonal, smooth and converge fast, they do not have a Fast Transform, which limits their utility.

## 6.4 Chebyshev polynomials

These are simply defined by

$$T_n(\cos \theta) = \cos n\theta.$$

It immediately follows

$$T_0(x) = 1,$$
$$T_1(x) = x,$$
$$T_2(x) = 2x^2 - 1,$$
$$T_3(x) = 4x^3 - 3x,$$
$$T_4(x) = 8x^4 - 8x^2 + 1.$$

The Chebyshev polynomials are orthogonal with the weight function $w(x) = 1/\sqrt{1 - x^2}$,

$$\int_{-1}^{1} T_m(x)T_n(x)w(x)\,dx = \begin{cases} 0 & \text{when } m \neq n, \\ \pi & \text{when } m = n = 0, \\ \frac{1}{2}\pi & \text{when } m = n \neq 0. \end{cases}$$

Chebyshev polynomials satisfy the ordinary differential equation

$$(1 - x^2)T_n'' - xT_n' + n^2 T_n = 0.$$

They satisfy useful recurrence relations

$$T_{n+1} = 2xT_n - T_{n-1},$$

$$T_n' = 2(n + 1)T_n + \frac{n + 1}{n - 1}T_{n-1}'.$$

From the definition of Chebyshev polynomials, all their maxima are $+1$ and all their minima are $-1$. As a result, an approximation to a function with a finite expansion in Chebyshev polynomials tends to be quite near the optimal polynomial, the polynomial which has the least maximum pointwise error.

## 6.5 Rates of convergence

If $f(\theta)$ is fully periodic and if $f(\theta)$ has $k$-derivatives, then integrating by parts $k$ times gives

$$\hat{f}_n = \frac{1}{2\pi} \int_0^{2\pi} f(\theta)e^{-in\theta}\,d\theta = \frac{1}{2\pi} \frac{i^k}{n^k} \int_0^{2\pi} f^{(k)}(\theta)e^{-in\theta}\,d\theta.$$

Thus the Fourier series converges rapidly with $\hat{f}_n = o(n^{-k})$ ($o()$ not $O()$ by the Reimann–Lebesgue lemma).

If $f^{(k)}$ has a discontinuity, then $\hat{f}_n = O(n^{-k-1})$.

If all the derivatives of $f$ are continuous, i.e. $f$ is $C^\infty$, then $\hat{f}_n$ decreases faster than any negative power of $n$, and typically decreases exponentially fast, like $e^{-Kn}$ for some constant $K$. For example, the fully periodic function

$$f(\theta) = \sum_{n=-\infty}^{\infty} \frac{1}{(\theta - 2\pi n)^2 + a^2}$$

has a transform

$$\hat{f}_n = \frac{\pi}{a} e^{-|n|a}.$$

Note that the convergence is controlled by the singularities of $f(\theta)$ in the complex $\theta$-plane.

## 6.6 Gibbs phenomenon

Discontinuous functions are not good for spectral representations. A discontinuity produces a $O(1/n)$ decrease in the spectral amplitudes, which means that the sum of spectral terms is not absolutely convergence, i.e. the convergence depends critically on the oscillating signs of the terms. Further there is a Gibbs phenomenon.

Figure 6.2 shows three finite Fourier series representations of a unit square wave, which takes values $+1$ in $(0, \pi)$ and $-1$ in $(\pi, 2\pi)$. There are discontinuities at $0$, $\pi$ and $2\pi$. The three curves in the figure have finite series including $N = 12$, 25 and 50 nonzero terms. Away from the discontinuities, the series convergences are pointwise to the square wave with errors $O(1/N)$. Near to the discontinuity, there is always an 18% overshoot, at a distance $1.5/N$ from the discontinuity.

One way to work with discontinuous functions is to write the function as a sum of discontinuities, whose positions and amplitudes have to be determined somehow, plus a smooth remainder. A spectral representation can be safely used for the smooth remainder, which should have reasonable convergence properties and no Gibbs phenomenon. However, it is probable that there are discontinuities in some derivatives of the remainder, which would still limit the convergence.

When working on a finite interval, it is necessary to consider whether or not the function is fully periodic in its value and all its derivatives. If it is

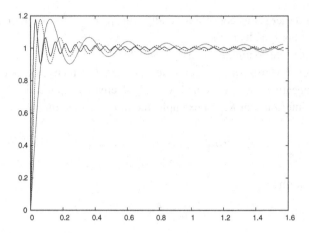

Figure 6.2 Gibbs phenomenon. Fourier series representations of the unity square wave, summing to 12, 25 and 50 terms.

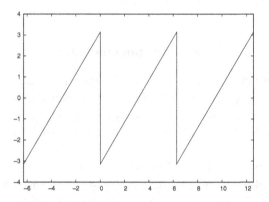

Figure 6.3 Hidden discontinuities at the ends of the interval $[0, 2\pi]$ of a non-fully periodic function.

fully periodic, then a Fourier series is appropriate. If not, then one should use a Chebyshev representation. The issue is that a Fourier series will be fully periodic, so must struggle to represent a function which is not.

Consider the function

$$f(\theta) = \theta - \pi \quad \text{over } (0, 2\pi).$$

The periodic extension of this function takes values

$$\theta + \pi \quad \text{in } (-2\pi, 0), \quad \text{and} \quad \theta - 3\pi \quad \text{in } (2\pi, 4\pi).$$

There are therefore hidden discontinuities at the ends of the intervals, at $\theta = -2\pi$, $0$, $2\pi$ and $4\pi$, see Figure 6.3 These hidden discontinuities lead to poor convergence and a Gibbs phenomenon.

How then does a representation by Chebyshev polynomials handle non-periodic functions and avoid the poor convergence and Gibbs phenomena? The key is the coordinate stretch $x = \cos\theta$. Consider the same linear function which takes different values at its ends

$$f(x) = x \quad \text{over } [-1, 1].$$

Viewed as a function of $\theta$ this function is

$$\tilde{f}(\theta) = \cos\theta \quad \text{over } [-\pi, 0].$$

The periodic extension of $\tilde{f}(\theta)$ is just $\cos\theta$ over all the $\theta$-real line, which is fully periodic. For a general nonperiodic function $f(x)$ over $[-1, 1]$, the Chebyshev representation uses the stretched variable $x = \cos\theta$ to produce a fully periodic $\tilde{f}(\theta) = f(\cos\theta)$ over $[-\pi, 0]$ and then takes an ordinary Fourier series of $\tilde{f}(\theta)$. (Note that the derivative $d\tilde{f}/d\theta = -\sin\theta f'(\cos\theta)$ vanishes at the ends of the $\theta$-interval $[-\pi, 0]$.)

## 6.7 Discrete Fourier Transform

Consider the Fourier series with modes $e^{in\theta}$ for $n = -M, \ldots, M$, i.e. odd $N = 2M + 1$ modes. Use $N$ equispaces collocations points to cover the interval $[0, 2\pi]$, i.e. collocations points at $\theta_j = 2\pi j/N$ for $j = 1, \ldots, N$.

While the coefficients $\hat{f}_n$ of the true Fourier series are found from an integral

$$\hat{f}_n = \frac{1}{\pi} \int_0^{2\pi} e^{-in\theta} f(\theta) \, d\theta,$$

the *Discrete* approximation $\tilde{f}_n$ is found by a sum over the collocations points

$$\tilde{f}_n = \frac{1}{N} \sum_{j=1}^{N} f(\theta_j) e^{-in\theta_j} \quad \text{for } n = -M, \ldots, M.$$

Note that because $\theta_j$ is a multiple of $2\pi/N$, then $e^{-i(n\pm N)\theta_j} = e^{-in\theta_j}$, so that

$$\tilde{f}_{n\pm N} \equiv \tilde{f}_n.$$

To find the Inverse Discrete Fourier Transform, one first considers an $N$th root of unity $\omega = e^{i2\pi/N}$. Because it is not unity and because $\omega^N - 1 = 0$, the sum of $N$ powers of the root of unity vanishes,

$$\sum_{j=-M}^{M} \omega^j = 0.$$

Then substituting the definition of $\tilde{f}_n$ and interchanging the order of summation

$$\sum_{n=-M}^{M} \tilde{f}_n e^{in\theta} = \sum_{j=1}^{N} f(\theta_j) \underbrace{\frac{1}{N} \sum_{n=-M}^{M} e^{in(\theta-\theta_j)}}.$$

If $\theta = \theta_k \neq \theta_j$ then the underbraced term is a sum of $N$ powers of a root of unity, and so the underbraced term vanishes. On the other hand if $\theta = \theta_j$, then the underbraced term is 1. Hence

$$\sum_{n=-M}^{M} \tilde{f}_n e^{in\theta} = f(\theta_j) \quad \text{if } \theta = \theta_j,$$

i.e. we recover the exact values at the collocations points.

The amplitudes $\hat{f}_n$ of the Continuous Fourier Transform are found from an integral, while the amplitudes $\tilde{f}_n$ of the Discrete Fourier Transform are found from a sum over values at equispaced points, and are thus numerical approximations of $\hat{f}_n$ with a small numerical error. It is important that the Inverse Discrete Fourier Transform, again a sum which is a numerical approximation to an integral, returns at the collocation points the exact values $f(\theta_i)$ of the original function, with the two numerical errors precisely cancelling.

## 6.8 Aliasing

Consider a function which oscillates at a high frequency, at a frequency outside the range $n = -M, \ldots, M$, e.g. $f(\theta) = e^{i(k+N)\theta}$. The Discrete Fourier Transform of this function has the erroneous result

$$\tilde{f}_k = \frac{1}{N} \sum_{j=1}^{N} f(\theta_j) e^{-ik\theta_j} = 1.$$

Thus oscillations outside the basic frequency range $n = -M, \ldots, M$ are erroneously mapped onto this range by subtracting appropriate multiples $N$. Figure 6.4 gives an illustration of this phenomenon, sometimes called the 'wagon-wheel' effect in which the spokes of wheels can appear to rotate backwards when viewed at a cinematic frame rate.

In nonlinear problems, high frequencies are generated through the sum and

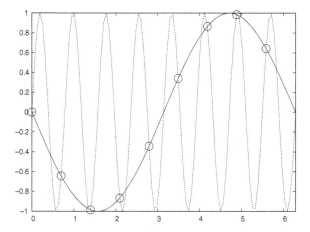

Figure 6.4 With $N = 9$ equi-spaced points, the high frequency $n = 8$ sinusoid takes the same values at the data points as a sinusoid with reduced frequency $n - N = -1$.

difference of frequencies of the components of the nonlinear term. It is necessary to avoid these correct high-frequency terms being erroneously interpreted as a low-frequency term by the aliasing phenomenon. There are several ways of *dealiasing* the problem, perhaps the simplest being the $\frac{2}{3}$-rule. In the $\frac{2}{3}$-rule, one starts with all the variables having frequencies in the range $[-M, M]$. Then one sets to zero the amplitudes of the modes outside the central $\frac{2}{3}$rds of the range, i.e. zero in $[-M, -\frac{2}{3}M]$ and $[\frac{2}{3}M, M]$. The product of two terms with frequencies from the truncated central range would then produce correct amplitudes with frequencies in the range $[-\frac{4}{3}M, \frac{4}{3}M]$. The correct amplitudes in the central $[-\frac{2}{3}M, \frac{2}{3}M]$ range are retained. The correct amplitudes in the range $[-\frac{4}{3}, -\frac{2}{3}M]$ would be erroneously interpreted as in the frequency range $[\frac{2}{3}M, \frac{4}{3}M]$ and so are set to zero, and similarly those in the range $[\frac{2}{3}M, \frac{4}{3}M]$. Hence the strategy is first to zero the amplitudes outside the central $\frac{2}{3}$rds frequency range, then form the product of two terms, and then again zero the amplitudes outside the central range. As one is going to zero the amplitudes outside the central $\frac{2}{3}$rds range, one need not calculate them in the first place. In order not to waste storage, one might better pretend that one started with a range $\frac{3}{2}$ larger which with the $\frac{2}{3}$rds reduction results in a fully employed store. Hence this method of dealiasing is sometimes called the $\frac{3}{2}$-rule.

In one dimension, the $\frac{2}{3}$rds rule throws away $\frac{1}{3}$ of the data. In two dimensions, $\frac{5}{9}$ths of the data are thrown away, and in three dimensions $\frac{19}{27}$ths, i.e. rather a lot.

## 6.9  Fast Fourier Transform (FFT)

The calculation to take the Discrete Fourier Transform

$$\tilde{f}_n = \sum_{j=1}^{N} f(\theta_j)\omega^{nj} \quad \text{for} \quad n = -\tfrac{N}{2}, \ldots, \tfrac{N}{2},$$

with $\theta_j = 2\pi j/N$ and $\omega = e^{i\theta_1}$, looks like $N$ amplitudes, each being the sum of $N$ terms, and hence a calculation of $N^2$ operations. This is expensive.

But one can split the sum into even and odd terms

$$\tilde{f}_n = \sum_{k=1}^{N/2} f(\theta_{2k})\omega_2^{nk} + \omega^{-n} \sum_{k=1}^{N/2} f(\theta_{2k-1})\omega_2^{nk}, \quad \text{with} \quad \omega_2 = \omega^2,$$

which looks like two separate DFTs on $\frac{N}{2}$ points, i.e. $2(\frac{N}{2})^2 = \frac{1}{2}N^2$ operations. If $N = 2^M$, one can repeat $M$ times the process of halving the number of operations, resulting in just $N\log_2 N$ operations.

Programming an FFT requires the identification of whether the index $j$ of the point $\theta_j$ is even or odd at each of the $M$ halving levels. This amounts to considering the binary representation of $j$. There are slick ways to program this issue, such as the Cooley–Tukey FFT algorithm. It is recommended that one does not write the code oneself, but instead calls one of the many well-written packages.

In two dimensions, one can use the Orszag speedup of partial sums to make considerable savings in summing a spectral representation. Consider

$$f(x_i, y_j) = \sum_{m=1}^{M} \sum_{n=1}^{N} a_{mn}\phi_m(x_i)\phi_n(y_j) \quad \text{for} \quad i = 1, \ldots, M, \quad j = 1, \ldots, N.$$

This looks like $MN$ results, one each for the different values of $i$ and $j$, each result being the sum over $MN$ terms, i.e. $M^2N^2$ total operations. However one can first form the partial sum

$$g_m(y_j) = \sum_{n=1}^{N} a_{mn}\phi_n(y_j).$$

There are $MN$ such partial sums, one for each of the different $m$ and $j$. Each partial sum is the sum over $N$ terms, i.e. $MN^2$ operations. Now one can use these partial sums in the final summation

$$f(x_i, y_j) = \sum_{m=1}^{M} g_m(y_j)\phi_m(x_i),$$

which is $MN$ results, each result being the sum of $M$ term. Hence by using the intermediate partial sums the total number of operations has been reduced from $M^2N^2$ to $M^2N + MN^2$.

In three dimensions, the sum of a spectral representation

$$f(x_i, y_j, z_k) = \sum_{l=1}^{L} \sum_{m=1}^{M} \sum_{n=1}^{N} a_{lmn} \phi_l(x_i) \phi_m(y_j) \phi_n(z_k)$$

$$\text{for} \quad i = 1, \ldots, L, \quad j = 1, \ldots, M, \quad k = 1, \ldots, N$$

looks like $LMN$ terms with each the sum of $LMN$ terms, i.e. $L^2M^2N^2$ operations. But using partial sums

$$g_{lm}(z_k) = \sum_{n=1}^{N} a_{lmn} \phi_n(z_k), \quad \text{and} \quad h_l(y_j, z_k) = \sum_{m=1}^{M} g_{lm}(z_k) \phi_m(y_j),$$

$$f(x_i, y_j, z_k) = \sum_{l=1}^{L} h_l(y_j, z_k) \phi_l(x_i)$$

one can reduce the total number of operations to $L^2MN + LM^2N + LMN^2$. This partial summation approach can be combined with the fast FFT summation in each one-dimensional part.

## 6.10 Differential matrix

To differentiate data with exponential accuracy one can use a spectral representation,

$$f(\theta_j) \stackrel{\text{transform}}{\longrightarrow} \tilde{f}_n \stackrel{\text{differentiate}}{\longrightarrow} \tilde{f}_n' = n\tilde{f}_n \stackrel{\text{transform}}{\longrightarrow} f'(\theta_i).$$

All the operations are linear operations, and hence the final answer is a linear sum of the input data, i.e.

$$f'(\theta_i) = D_{ij} f(\theta_j),$$

with differentiation matrix $D$. The $N^2$ operations in evaluating the derivative this way at $N$ points can be reduced to $N \log_2 N$ operations by an FFT factorisation. Note that the matrix for the second derivative is not the square of the matrix for the first derivative.

One can see a simple pattern in the accuracy of different estimates of the derivative. With finite differences, using two data points can give second-order accuracy, i.e. errors $O(N^{-2})$; using four data points can give fourth-order accuracy, i.e. errors $O(N^{-4})$; and so using all $N$ data points gives exponential accuracy, i.e. $O(N^{-N})$ errors.

## 6.11 Navier–Stokes

Taking the Fourier transform of the Navier–Stokes equations

$$\nabla \cdot \mathbf{u} = 0,$$

$$\frac{\partial \mathbf{u}}{\partial t} + \mathbf{u} \cdot \nabla \mathbf{u} = -\nabla p + \nu \nabla^2 \mathbf{u},$$

one obtains

$$\mathbf{k} \cdot \hat{\mathbf{u}} = 0,$$

$$\frac{\partial \hat{\mathbf{u}}}{\partial t} + \widehat{\mathbf{u} \cdot \nabla \mathbf{u}} = -i\mathbf{k}\hat{p} - \nu k^2 \hat{\mathbf{u}}.$$

The pressure is found by taking a scalar product of the latter with $\mathbf{k}$ and using the former,

$$\hat{p} = \frac{\mathbf{k} \cdot \widehat{\mathbf{u} \cdot \nabla \mathbf{u}}}{-ik^2}.$$

Solving the Poisson problem is thus trivial. Substituting this pressure back into the transform of the momentum equation,

$$\frac{\partial \hat{\mathbf{u}}}{\partial t} = -\left(\mathbf{I} - \frac{\mathbf{k}\mathbf{k}}{k^2}\right) \cdot \widehat{\mathbf{u} \cdot \nabla \mathbf{u}} - \nu k^2 \hat{\mathbf{u}}.$$

The above solution with Fourier transforms is appropriate to a fully periodic problem which would not have boundary conditions. When there are finite boundaries, one uses a Chebyshev spectral representation. If the boundary conditions are that a function vanishes there, one can automatically satisfy these conditions by using as the basis functions combinations of Chebyshev polynomials

$$\phi_{2n} = T_{2n} - T_0 \quad \text{and} \quad \phi_{2n-1} = T_{2n-1} - T_1.$$

If the boundary conditions are that a function does not vanish there, then an alternative 'tau' method can be employed of imposing the boundary condition

$$\sum_{n=1}^{N} \hat{f}_n T_n(x) = \text{boundary value at } x = \pm 1.$$

One part of the Navier–Stokes equations is the diffusion equation

$$u_t = D u_{xx}.$$

The explicit forward time-stepping algorithm has a stability criterion $\Delta t < 0.25 \Delta x^2 / D$. Now a Chebyshev spectral representation of $N$ terms would use collocation points which are equally spaced in $\theta$, which give a crowding of

Table 6.1 *Methods bridging local and global properties.*

| Local | | | | Global |
|---|---|---|---|---|
| | *Finite elements* | FE $h^p$ | | |
| *Finite differences* | | | | *Spectral* |
| point data | | | | whole interval |
| | *Splines* | *Wavelets* | | |
| | global points | local waves | | |

the points $x = \cos\theta$ near to the boundaries, where the smallest spacing is $\Delta x = \pi^2/2N^2$. The stability criterion would then be $\Delta t < \pi^4/16DN^4$, which is quite a limitation. On the other hand, one does not need so many terms in a spectral representation compared with finite differences, perhaps only $3/\sqrt{\delta}$ terms to resolve a boundary layer of thickness $\delta$.

## 6.12 Bridging the gap

Returning to the issue of local and global, we can now survey the different methods of discretisation, see Table 6.1. *Finite differences* are strictly local, representing the data by values at fixed spatial points. A *Spectral* representation, on the other hand, is strictly global with each basis function being nonzero across the whole spatial range. *Finite elements* are nearer local than global, with the basis functions nonzero is small patches. However moving towards global, there are some so-called $h^p$ finite elements which have a $p$ high-order polynomial representation within fewer larger patches. In Part III in §11.1.1 we will discuss *splines*, in which functions are represented across the whole range with piecewise low-order polynomials. Splines have therefore some global aspects while being nearer to local in nature. Finally *wavelets*, which will also be discussed in Part III in Chapter 17, represent functions as sums of localised waves with different wavelengths, and these have some local aspects while being nearer to global in nature.

## Further Reading

*Spectral methods in fluid dynamics* by M. Y. Hussaini and T. Y. Zang in Annu. Rev. Fluid Mech. (1987) **19**, 339–367.

# 7

# Time integration

When looking for an algorithm to integrate an equation forward in time, there are three important issues to consider:

- accuracy, and how does it depend on the size of the time-step $\Delta t$,
- cost, both the CPU run-time and the memory storage required, and
- stability, a subtle issue to be discussed shortly.

For this chapter we consider the integration in time independently of the spatial discretisation. The spatial dependency in a partial differential equation may be rendered by either finite differences, finite elements or a spectral representation to a coupled ordinary differential equation for the time evolution of the large vector $u(t)$ of the values or amplitudes in the spatial discretisation, i.e. the problems is reduced to the form

$$u_t = F(u, t).$$

The time evolution is then treated by a 'black-box' time-integrator which is unaware of the spatial discretisation. This is typical of the approach that must be used with finite elements and spectral methods. With finite differences, it is possible to exploit the known method of spatial representation to create superior time-integrators, particularly for hyperbolic partial differential equations. The later Chapter 10 in Part III on special topics will consider hyperbolic equations. For this chapter, we consider integration in time ignorant of the nature of the spatial discretisation.

## 7.1 Stability

There are three classes of algorithms with regard to stability.

- Unstable algorithms. These are bad!
  The numerical solution blows up in time, whatever the value of the time-step $\Delta t$, and usually blows up rapidly, and often oscillates as it blows up.
- Conditionally stable algorithms. This is the normal type.
  The algorithms are stable so long as $\Delta t$ is smaller than some limit.
- Unconditionally stable algorithms. These algorithms are slightly dangerous. They are stable for all values of $\Delta t$, and this can be misleading because the 'stable' solution can be very inaccurate at larger $\Delta t$.

When applied to algorithms, the word 'stable' can have at least three shades of meaning.

- The numerical solution decays, even if the physics does not.
  This is not good.
- The numerical solution does not blow up for all time, i.e. remains bounded.
  This is a little better.
- The numerical solution blows up but only slowly, and for any fixed time $t$ the numerical solution at that time converges as $\Delta t \to 0$.
  This is best.

In the latter case, it is highly probable that ever smaller values of $\Delta t$ must be used when taking the solution to longer and longer times.

A suitably stable algorithm is obviously desirable. Stability also brings the benefit of getting the correct solution, according the to *Lax equivalence theorem*. This theorem states that a consistent algorithm, meaning one in which the local errors for a single step decrease when $\Delta t$ tends to zero, converges to the correct solution if and only if the algorithm is stable.

In a large coupled system of equations, there can be many modes with a wide range of different time-scales. This raises the issue of the *stiffness* of the system of equations, which has a major impact on the stability of algorithms. Consider a system of equations

$$u_t = F(u, t).$$

We linearise the problem to consider small disturbances $\delta u(t)$ about a solution $u_0$ and then freeze the coefficients at time $t = t_0$. (This latter step can occasionally lead to misleading conclusions.)

$$\delta u_t = F'(u_0, t_0)\delta u.$$

One can then find the eigenvalues $\lambda$ of the linearisation Jacobian matrix $F'$. If the modulus of the largest eigenvalue is very much larger than the modulus of the smallest,

$$|\lambda_{max}| \gg |\lambda_{min}|,$$

the system is called *stiff*. Stability is controlled by the largest eigenvalue, with typically a condition

$$\Delta t < \text{const}/|\lambda_{max}|.$$

This largest eigenvalue may represent some boring behaviour of fast modes on small scales, e.g. modes that just decay, whereas one is really interested in much slower modes on longer scales. In such a situation, it may be acceptable to use an unconditionally stable algorithm with a $\Delta t$ much bigger than the above limit, which then would probably render the boring fast modes inaccurately, but choose $\Delta t < \text{const}/|\lambda_{min}|$, in order to render accurately the interesting slow modes.

## 7.2 Forward Euler

This algorithm is explicit and only first-order accurate in $\Delta t$. Explicit means that the algorithm gives an explicit formula for $u^{n+1}$ rather than an implicit expression that must be solved. First-order means that there are errors $O(\Delta t)$ in the numerical approximation to the correct answer at some fixed time, say $t = 1$. For the simple ordinary differential equation

$$u_t = \lambda u,$$

the algorithm is

$$\frac{u^{n+1} - u^n}{\Delta t} = \lambda u^n,$$

with $n$ superscripts for the numerical approximation to $u(t = n\Delta t)$. Solving the algorithm, we have

$$u^n = (1 + \lambda \Delta t)^n u^0.$$

Now the number of time-steps to reach a finite time $t$ is $n = t/\Delta t$, so as $\Delta t \to 0$,

$$u^n \to e^{\lambda t} u^0,$$

which is the correct solution of the original differential equation.

In the *case of $\lambda$ real and negative*, the algorithm is stable if

$$\Delta t < 2/|\lambda|.$$

If $\Delta t$ is larger, the numerical solution blows up. Once $\Delta t > 1/|\lambda|$, the numerical solution oscillates in sign each time-step, which the correct solution of the differential equation does not.

In the *case of $\lambda$ purely imaginary*, the amplitude of the numerical solution increases by a factor of

$$|1 + \lambda\Delta| = \left(1 + |\lambda|^2\Delta t^2\right)^{1/2}$$

at each time-step. After $n = t/\Delta t$ time-steps, the amplitude is

$$\left(1 + |\lambda|^2\Delta t^2\right)^{\frac{t}{2\Delta t}} \quad \rightarrow \quad e^{\frac{1}{2}|\lambda|^2\Delta t\, t} \quad \text{as } \Delta t \rightarrow 0,$$

i.e. the amplitude does blow up, but not too much, say by $\epsilon$, if

$$\Delta t < \frac{2\ln\epsilon}{|\lambda|^2 t}.$$

## 7.3 Backward Euler

This algorithm is implicit and very stable, but only first-order accurate. For again $u_t = \lambda u$, the algorithm is

$$\frac{u^{n+1} - u^n}{\Delta t} = \lambda u^{n+1}.$$

The algorithm is implicit because the right-hand side has the unknown $u^{n+1}$. For the linear differential equation being considered, it is no challenge to solve the implicit expression for

$$u^n = \left(\frac{1}{1 - \lambda\Delta t}\right)^n u_0.$$

For a nonlinear problem, solving the nonlinear implicit relation may require considerable computer costs.

The Backward Euler algorithm is stable outside the circle on the complex $\lambda$-plane

$$|1 - \lambda\Delta t| = 1,$$

and unstable inside the circle. However it is unwise to use a large $\Delta t$ because the numerical approximation can become very inaccurate, e.g. when $\lambda$ is real and negative setting $\Delta t = 1/|\lambda|$ gives

$$u(t) \sim e^{\lambda t \ln 2} \quad \text{instead of the correct} \quad e^{\lambda t}.$$

## 7.4 Midpoint Euler

This algorithm is explicit and second-order accurate, i.e. the numerical solution deviates from the correct answer by $O(\Delta t^2)$ at any fixed time. This algorithm has the merit that it is simple to recode the Forward Euler to this algorithm, so upgrading a first-order scheme to second-order. Previous algorithms have been written for a linear differential equation, with the generalisation to a nonlinear equation straightforward. Here we consider the general nonlinear problem $u_t = F(u, t)$. The midpoint algorithm is

$$\frac{u^* - u^n}{\frac{1}{2}\Delta t} = F(u^n, t_n), \qquad \frac{u^{n+1} - u^n}{\Delta t} = F\left(u^*, t_{n+\frac{1}{2}}\right).$$

For the linearised equation $u_t = \lambda u$, the midpoint algorithm has the same stability behaviour as the Forward Euler algorithm for both real negative $\lambda$ and for purely imaginary $\lambda$.

## 7.5 Crank–Nicolson

The Crank–Nicolson algorithm is implicit and second-order accurate. It has the advantage of only evaluating the derivative only once per time-step which partially offsets the disadvantage of being implicit. For the equation $u_t = \lambda u$ again, the algorithm is

$$\frac{u^{n+1} - u^n}{\Delta t} = \lambda \frac{u^{n+1} + u^n}{2},$$

i.e. it uses central differencing about the half time-level $t = (n + \frac{1}{2})\Delta t$. Note that the algorithm is implicit in that the right-hand side involves the unknown $u^{n+1}$. For the linear differential equation being considered, this implicitness is not a problem. Solving we have

$$u^n = \left(\frac{1 + \frac{1}{2}\lambda\Delta t}{1 - \frac{1}{2}\lambda\Delta t}\right)^n u^0.$$

In the *case of Re($\lambda$)* $< 0$, the denominator is larger than the numerator for all values of $\Delta t$. Hence the algorithm is stable.

In the *case of $\lambda$ purely imaginary*, the numerator and denominator have the same modulus so that the amplitude remains correctly constant for all values of $\Delta t$. Hence the algorithm is stable. The phase of the oscillation, however, drifts slowly through the $O(\Delta t^2)$ errors.

## 7.6 Leapfrog

This algorithm is explicit, evaluates the derivative only once per time-step and is second-order accurate, but is sometimes hopelessly unstable. For $u_t = \lambda u$, the algorithm is central differenced around time-level $n$ by using information from levels $n + 1$ and $n - 1$,

$$\frac{u^{n+1} - u^{n-1}}{2\Delta t} = \lambda u^n.$$

This is a two-term recurrence relation,

$$u^{n+1} - 2\lambda \Delta t u^n - u^{n-1} = 0,$$

which has solutions $u^n = A\theta_+^n + B\theta_-^n$ with $\theta_\pm = \lambda \Delta t \pm \sqrt{1 + \lambda^2 \Delta t^2}$. So as $\Delta t \to 0$,

$$u^n \sim e^{\lambda n \Delta t} + \epsilon(-1)^n e^{-\lambda n \Delta t}.$$

The first term is the correct solution of the differential equation, while the second term is a spurious solution which changes sign each time-step and blows up if $Re(\lambda) < 0$. Hence for stable ordinary differential equations the algorithm is unstable for all $\Delta t$.

The Leapfrog algorithm is however stable for purely imaginary $\lambda$ and $\Delta t < 1/|\lambda|$.

## 7.7 Runge–Kutta

There are a number of Runge–Kutta methods for time-integrating differential equations, some explicit and some implicit, methods with different orders of accuracy, methods involving different numbers of function calls and methods of varying stability. The Euler Midpoint method can be consider a second-order accurate two-stage Runge–Kutta method.

Perhaps the best-known and most commonly used Runge–Kutta method is the explicit fourth-order algorithm. For a general nonlinear equation $u_t = F(u, t)$, the algorithm is

$$du^1 = \Delta t F\left(u^n, t^n\right),$$
$$du^2 = \Delta t F\left(u^n + \tfrac{1}{2} du^1, t^n + \tfrac{1}{2}\Delta t\right),$$
$$du^3 = \Delta t F\left(u^n + \tfrac{1}{2} du^2, t^n + \tfrac{1}{2}\Delta t\right),$$
$$du^4 = \Delta t F\left(u^n + du^3, t^n + \Delta t\right),$$
$$u^{n+1} = u^n + \tfrac{1}{6}\left(du^1 + 2du^2 + 2du^3 + du^4\right).$$

Note that this four-stage method requires the derivative $F(u, t)$ to be evaluated four times. These function calls can be expensive, for example if the time derivative hides a complex spatial discretisation with perhaps a Poisson problem to invert. One advantage of Runge–Kutta methods is that the time-step $\Delta t$ can be adjusted every time-step in order to maintain good accuracy. In general, the Runge–Kutta algorithms have good stability, requiring something like $\Delta t < 3/|\lambda_{max}|$.

Knowing that the above Runge–Kutta algorithm is fourth-order accurate, one can compare making one time-step of $2\Delta t$ with making two steps of $\Delta t$ in order to estimate the error. Let the additional error of making one time-step of $\Delta t$ from $u^n$ to $u^{n+1}$ be $B\Delta t^5$. Then the additional error of making two steps from $u^n$ to $u^{n+2}$ is $2B\Delta t^5$, while the error of making a single step of $2\Delta t$ from $u^n$ to $u^*$ is $32B\Delta t^5$. Hence the error in a single time-step of $\Delta t$ is estimated as

$$\frac{1}{30}\left(u^* - u^{n+2}\right).$$

If this error is larger than some preset tolerance, then $\Delta t$ can be reduced appropriately and the integration restarted from $u^n$. Subtracting the above estimate of the error from $u^{n+2}$ yields a corrected answer

$$\frac{16}{15}u^{n+2} - \frac{1}{15}u^*,$$

with errors of now $O(\Delta t^6)$ for a single step.

There are implicit Runge–Kutta schemes such as the Hammer–Hollingsworth algorithm

$$du^1 = \Delta t F\left(u^n + \tfrac{1}{4}du^1 + \left(\tfrac{1}{4} - \tfrac{\sqrt{3}}{6}\right)du^2, t^n + \left(\tfrac{1}{2} - \tfrac{\sqrt{3}}{6}\right)\Delta t\right),$$

$$du^2 = \Delta t F\left(u^n + \left(\tfrac{1}{4} + \tfrac{\sqrt{3}}{6}\right)du^1 + \tfrac{1}{4}du^2, t^n + \left(\tfrac{1}{2} + \tfrac{\sqrt{3}}{6}\right)\Delta t\right),$$

$$u^{n+1} = u^n + \tfrac{1}{2}du^1 + \tfrac{1}{2}du^2.$$

This algorithm is fourth-order accurate. It is stable for all $\Delta t$ if $Re(\lambda) \leq 0$, although not accurate if $\Delta t$ is too large. While the algorithm is a two-stage Runge–Kutta, many more than two function evaluations are likely to be required in an iterative solution of the implicit expressions if the problem is nonlinear.

## 7.8 Multistep methods

Multistep methods use information from several previous steps. A polynomial is fitted to the derivative at recent steps. This polynomial is used to predict the variation of the derivative in the current time interval, so avoiding extra function evaluations within the interval. Integrating the varying derivative, one predicts the value at the end of the current interval.

Adams–Bashforth algorithms are explicit. The following example is third-order accurate

$$u^{n+1} = u^n + \frac{\Delta t}{12} \left(23F_n - 16F_{n-1} + 5F_{n-2}\right).$$

Adams–Moulton algorithms are implicit. The following example is fourth-order accurate

$$u^{n+1} = u^n + \frac{\Delta t}{24} \left(9F_{n+1} + 19F_n - 5F_{n-1} + F_{n-2}\right).$$

These algorithms are stable for $\Delta t \lesssim 1/|\lambda|$. Note that both use a single new function evaluation per step, which is good. Note that they are difficult to start from the initial conditions, and it is difficult to change step size $\Delta t$ for them, which are both bad. To start the algorithms from the initial conditions, one must use another method, such as Runge–Kutta, to make the first two or three steps. Alternatively one can analytically generate the first few points by a suitably high-order Taylor series expansion. To change the step size, one must fit a suitably high-order polynomial through recent points in order to interpolate to the new recent points.

To solve the implicit expression for $u^{n+1}$ in the fourth-order Adams–Moulton algorithm, one can use to sufficient accuracy the third-order estimate of $u^{n+1}$ from the Adams–Bashforth algorithm. This combination is called a Predictor-Corrector scheme. This however means that two function evaluations are used for each step.

## 7.9 Symplectic integrators

Hamiltonian systems of differential equations

$$\dot{p}_i = -\frac{\partial H}{\partial q_i}, \qquad \dot{q}_i = \frac{\partial H}{\partial p_i},$$

which do not explicitly depend on time, i.e. $H(p, q)$ is independent of $t$, conserve energy, the value of the Hamiltonian $H$. They also preserve volumes in the $pq$-phase-space of trajectories of solutions. Symplectic integrators have the same conservation properties for a numerical approximation to the Hamiltonian $H^{\text{num}}(\Delta t)$. Note one must therefore keep $\Delta t$ fixed. This is important for integration to long times, when any loss of energy would dramatically change the solution.

An example of a symplectic integrator is the Störmer–Verlet algorithm, a sort of leapfrog, commonly used in molecular dynamics simulations of $m\ddot{r} = F(r)$.

$$p^{n+\frac{1}{2}} = p^n + \tfrac{1}{2}\Delta t F(r^n),$$

$$r^{n+1} = r^n + \Delta t \tfrac{1}{m} p^{n+\frac{1}{2}},$$

$$p^{n+1} = p^{n+\frac{1}{2}} + \tfrac{1}{2}\Delta t F\left(r^{n+1}\right).$$

At each of the three ministeps, one variable of $p$ and $r$ is changed while holding constant the other variable fixed at its value at the start of the ministep. This means that each ministep will transform an initial small rectangle in $pq$-phase-space into a parallelogram of the same area. Hence the combination of the three ministeps conserves volume in phase-space.

## 7.10 Navier–Stokes

This section combines several ideas to construct a stable second-order accurate algorithm to time-step the Navier–Stokes equations. A split time-step is used, from $u^n$ to $u^*$ to $u^{n+1}$, with the second part designed to project the solution so that it obeys exactly the solenoidal constraint $\nabla \cdot \mathbf{u} = 0$.

The first part of the split time-step combines different methods for different terms, central differencing around the time-level $n + \frac{1}{2}$,

$$\frac{\mathbf{u}^* - \mathbf{u}^n}{\Delta t} = -(\mathbf{u} \cdot \nabla \mathbf{u})^{n+\frac{1}{2}} - \nabla p^{n-\frac{1}{2}} + \nu \nabla^2 \left(\frac{\mathbf{u}^* + \mathbf{u}^n}{2}\right).$$

The diffusion term, last on the right-hand side, is tackled implicitly, which gives good stability at boring fine length scales and second-order accuracy. The advection term, first on the right-hand side, can be obtained explicitly with third-order accuracy by a safe Adams–Bashforth method

$$(\mathbf{u} \cdot \nabla \mathbf{u})^{n+\frac{1}{2}} = \frac{1}{12}\left(23\,(\mathbf{u} \cdot \nabla \mathbf{u})^n - 16\,(\mathbf{u} \cdot \nabla \mathbf{u})^{n-1} + 5\,(\mathbf{u} \cdot \nabla \mathbf{u})^{n-2}\right).$$

The pressure is tackled in this first part of the split time-step by a so-called update method, which uses the currently available estimate of the pressure gradient from the last time-step. Boundary conditions must be imposed on $u^*$. The normal component of $u^*$ is handled as described at the end of §3.6. The tangential component is roughly set to the given tangential component $u_{\text{tang}}^{\text{BC}}$, but not quite; see below.

The second part of the split time-step is a projection onto the solenoidal constraint $\nabla \cdot \mathbf{u}^{n+1} = 0$.

$$\mathbf{u}^{n+1} = \mathbf{u}^* + \Delta t \nabla \phi^{n+1}$$

with

$$\nabla^2 \phi^{n+1} = -\nabla \cdot u^*/\Delta t.$$

A normal boundary condition is imposed,

$$\Delta t \frac{\partial \phi^{n+1}}{\partial n} = u_n^{BC} - u_n^*.$$

With this $\phi^{n+1}$, the velocity is then updated from $u^*$ to $u^{n+1}$, and the pressure is updated by

$$p^{n+\frac{1}{2}} = p^{n-\frac{1}{2}} - \phi^{n+1} + \tfrac{1}{2}\nu\Delta t \nabla^2 \phi^{n+1}.$$

Now the update of the velocity by $\Delta t \nabla \phi^{n+1}$ will have a small tangential component along the boundary. In order that the final $u^{n+1}$ satisfies that tangential boundary condition, it is necessary to anticipate this update by imposing on $u^*$ the tangential boundary condition

$$u_{\text{tang}}^* = u_{\text{tang}}^{BC} - \Delta t \nabla \phi^n.$$

While the update should be with $\phi^{n+1}$ rather than the $\phi^n$ in the above, the resulting error in the tangential boundary condition will be only $O(\Delta t^2)$.

# 8

# Linear algebra

HEALTH WARNING. Do not do it. Do not even contemplate it. Do not code up any of the linear algebra algorithms described in the chapter. Instead use the excellent free packages, such as LAPACK. These packages are professionally written. They are safe: if there is a problem such as being asked to solve an insoluble problem, they will fail gracefully with a helpful remark. They are also fast, designed to be efficient, often using hardware-dependent subroutines to streamline the low-level calculations.

The purpose of the chapter is therefore not to provide codable algorithms but to describe the large variety of algorithms for different circumstances so that the correct package can be selected. The chapter is a brief review of a large subject. There are many good, long textbooks which go into much greater detail, which may not be what a user of packages needs to know.

While fluid mechanics is nonlinear, there are many steps to finding the full solution which are linear. Thus the Poisson problem for finding the pressure field is a large linear algebra problem which commonly consumes most of the CPU computing time. Studying the linear stability is again a linear algebra problem of finding eigenvalues.

There are two basic questions described in this chapter.

- 'Matrix inversion', i.e. solving $Ax = b$ to find the vector $x$ when given the matrix $A$ and vector $b$.

- Eigenvalues, i.e. finding the eigenvalues $\lambda$ and eigenvectors $e$ in $Ae = \lambda e$ for a given $A$.

Within these two basic questions there are variations, such as a least-square solution of linear equations, singular value decomposition and the generalised eigenvalue problem $Ae = \lambda Be$.

The matrices $A$ involved can have different properties, such as

- dense or sparse,
- symmetric, positive definite, banded, . . . .

These properties will have a major impact on the method and speed of solution, so should be exploited in the selection of the appropriate package.

This chapter splits into two halves, the first half on solving linear equations and the second half on finding eigenvalues. Three approaches to solving linear equations are described in separate sections: Gaussian elimination and its refinements, *QR* decompositions by three different techniques and the conjugate gradients method. Similarly three approaches to finding eigenvalues are described in separate sections: the power method, the Jacobi method and the main method.

## 8.1 LAPACK

The LAPACK package is free and can be downloaded from the web. It replaces the earlier packages LINPACK and EISPACK, being more appropriate to modern CPUs with large caches. It uses the BLAS package of low-level functions which are tuned for the particular CPU.

The LAPACK organisation (netlib.org) has a search engine on the web to help select the correct routine, which can then be downloaded along with its dependencies. To find the correct route, one must specify the following.

- The problem: linear equations, least-squares solution, eigenvalue, singular decomposition, generalised eigenvalue; LU, QR, Cholesky and Schur decomposition.
- The precision: single or double precision, real or complex variables.
- The type of matrix: diagonal, tridiagonal or banded; real symmetric, Hermitian or Unitary; symmetric positive definite; Hessenberg, upper Hessenberg.

The naming of the package's subroutines is initially obscure. To solve a system of linear equations one can use the routine

$$sgesv(N, NRHS, A, LDA, IPIV, B, LDB, INFO).$$

In the name *sgesv*, the first *s* stands for real single precision, with alternatives *d* for double, *c* for complex and *z* for double precision complex. The middle *ge* stands for a general matrix, with many alternatives. The final *sv* stands for solving linear equations. The routine solves not just $Ax = b$ for one input vector $b$, but solves the equation for *NRHS* input vectors, i.e. it solves $AX = B$ for an $N \times N$ matrix $A$ and $N \times NRHS$ matrices $X$ and $B$. One inputs into the routine the dimensions $N$ and $NRHS$ and matrices $A$ and $B$ and sets $LDA = LDB = N$. The routine outputs the solution $X$ in place of $B$, the LU decomposition of $A$ in place of $A$, the row-pivot permutation vector in $IPIV$ and important information in the flag $INFO$. If the operation was successful, then $INFO = 0$. If $INFO = -i < 0$, then an illegal argument was passed to the routine in location $i$. If $INFO = i$, then the matrix was singular at row $i$.

To solve the generalised eigenvalue problem $Ae = \lambda Be$ one can use the routine

$$zggev(\dots, N, A, \dots, B, \dots, ALPHA, BETA, \dots, INFO).$$

The eigenvalues are delivered as ratios $ALPHA(j)/BETA(j)$ in case some $BETA(j)$ vanish. The routine offers options of finding the left and right eigenvectors via the arguments not displayed above.

## 8.2 Gaussian elimination

This method solves the system of linear equations

$$
\begin{aligned}
a_{11}x_1 &+ a_{12}x_2 &+ \cdots &+ a_{1n}x_n &= b_1, \\
a_{21}x_1 &+ a_{22}x_2 &+ \cdots &+ a_{2n}x_n &= b_2, \\
&\vdots & & &\vdots \\
a_{n1}x_1 &+ a_{n2}x_2 &+ \cdots &+ a_{nn}x_n &= b_n,
\end{aligned}
$$

to find the unknown vector $x$ given the matrix $A$ and vector $b$.

One starts by dividing the first equation by $a_{11}$, so that the coefficient of $x_1$ becomes 1.

Next one subtracts from the $k$th row the now normalised first equation multiplied by $a_{k1}$ for $k \geq 2$. As a result of this subtraction, the coefficient $a_{k1}$ of $x_1$ in the $k$th row becomes 0.

These two steps are now repeated on the $(n-1) \times (n-1)$ subsystem of equations 2 to $n$, i.e. omitting the first row and first column. This results in the coefficient $a_{22}$ of $x_2$ in equation 2 becoming 1 and the coefficients $a_{2k}$ of $x_2$ in the $k$th row becoming 0 for $k \geq 3$.

This process of two steps is repeated on the sequence of the ever smaller subsystems, until after $n$ applications one arrives at the $1 \times 1$ subsystem for $x_n$ alone.

The matrix $A$ has now been reduced to an upper triangular form with elements of different values

$$
\begin{aligned}
x_1 &+ a'_{12}x_2 &+ \cdots &+ a'_{1n}x_n &= b'_1 \\
0 &+ x_2 &+ \cdots &+ a'_{2n}x_n &= b'_2 \\
&\phantom{+}\vdots & & &\phantom{=}\vdots \\
0 &+ 0 &+ \cdots &+ x_n &= b'_n.
\end{aligned}
$$

Finally, one can 'back-solve', starting at the last $n$th equation and winding back one equation at a time to reach finally the first equation.

$$
\begin{aligned}
x_n &= b'_n & \text{gives} \quad & x_n, \\
x_{n-1} + a'_{n-1\,n}x_n &= b'_{n-1} & \text{gives} \quad & x_{n-1}, \\
&\phantom{=}\vdots & & \\
& & \text{gives} \quad & x_1.
\end{aligned}
$$

For a dense matrix, the Gaussian elimination method requires $\frac{1}{3}n^3$ multiplications to solve for the vector $x$. If the matrix is tridiagonal and one skips nul calculations, then only $3n$ multiplications are required to find $b$. Hence spareness in the matrix can greatly change the time required.

### 8.2.1 Pivoting

There is a problem with simple Gaussian elimination if at step $k$ the top-left element vanishes, $a_{kk} = 0$. Now the order of the equations on the list to be solved has no effect on the solution vector $x$. Hence if $a_{kk}$ vanishes, one can swap the $k$th equation with an equation below that does not have a vanishing coefficient of $x_k$. This swapping of rows is called 'partial pivoting'. One searches rows $j = k$ to $n$ and finds the row $j$ with the largest $|a_{kj}|$. Rows, i.e. equations, $j$ and $k$ are then swapped. If all the elements $a_{kj}$ vanish for $j = k$ to $n$, then the matrix is singular, and one should abandon solving the system of linear equations, because either there exists no solution or there exists an infinite number of solutions. In fact the swap is not implemented by rewriting elements of the matrix $A$, especially when this a very big matrix, but instead one creates a swap table of rows, so that when the $k$th row is needed one actually uses through the table the $j$th row, and uses the $k$th row in place of the $j$th,

Swapping rows is partial pivoting. Full pivoting swaps rows and columns, but is rarely better.

## 8.2.2 LU decomposition

This is just a rephrasing of Gaussian elimination. The idea is to write the matrix $A$ as the product of a lower triangular matrix $L$ and an upper triangular matrix $U$

$$A = LU, \quad \text{with} \quad L = \begin{pmatrix} 1 & 0 & 0 & 0 \\ * & 1 & 0 & 0 \\ * & * & 1 & 0 \\ * & * & * & 1 \end{pmatrix}, \quad U = \begin{pmatrix} * & * & * & * \\ 0 & * & * & * \\ 0 & 0 & * & * \\ 0 & 0 & 0 & * \end{pmatrix}.$$

Following Gaussian elimination, one moves through a sequence of submatrices of decreasing size formed by progressively dropping rows from the top and columns from the left. At each step, one first deposits the current top row of the submatrix into $U$. Then the factors used to multiply the current top row when subtracting from rows below are recorded in $L$. Finally those subtractions are made from the rows remaining in $A$.

For $k = 1 \rightarrow n$:
  for $j = k \rightarrow n$: $u_{kj} = a_{kj}$,
  for $i = k \rightarrow n$: $\ell_{ik} = a_{ik}/a_{kk}$,
  for $i = k + 1 \rightarrow n$ and for $j = k + 1 \rightarrow n$: $a_{ij} \leftarrow a_{ij} - \ell_{ik}u_{kj}$.

Having found the $LU$ decomposition of $A$, one solves the linear equations $Ax = LUx = b$ as follows. First a forward sweep to find $y$ in $Ly = b$

$$\ell_{11}y_1 \qquad\qquad = b_1 \text{ gives } \quad y_1,$$
$$\ell_{21}y_1 + \ell_{22}y_2 = b_1 \text{ gives } \quad y_2,$$
$$\vdots$$
$$\text{gives } \quad y_n,$$

and then a backward sweep to find $x$ in $Ux = y$

$$u_{nn}x_n \qquad\qquad = y_n \qquad \text{gives} \qquad x_n,$$
$$u_{n-1\,n-1}x_{n-1} + u_{n-1\,n}x_n = y_{n-1} \quad \text{gives} \qquad x_{n-1},$$
$$\vdots$$
$$\text{gives} \qquad x_1.$$

Finding the $LU$ decomposition is $O(n^3)$ operations, while solving $LUx = b$ for a new $b$, is only $O(n^2)$.

If the matrix $A$ is symmetric, then the upper triangular matrix $U$ is related through the symmetry to the lower triangular matrix $L$ via a diagonal matrix $D, A = LDL^T$.

If the matrix $A$ is symmetric and positive definite, then the elements of the diagonal $D$ are positive and so have real square roots, so that a diagonal square root matrix $D^{1/2}$ can be defined. Absorbing one square root into $L$ and one into $L^T$, we have the Cholesky decomposition, $A = BB^T$ with $B = LD^{1/2}$.

If the matrix $A$ is tridiagonal, then the lower triangular matrix $L$ has nonzero just the diagonal and the elements one below the diagonal, while the upper triangular matrix $U$ has nonzero just the diagonal and the elements one above the diagonal.

Finally, making the $LU$ decomposition gives a fast way of evaluating the determinant of the matrix $A$ as the product of the diagonal elements of $U$, $\det A = \Pi_i u_{ii}$. This method of evaluating the determinant takes $n^3$ operations compared with $n!$ following the classical definition.

### 8.2.3 Errors

A small $\epsilon$ error in the input vector $b$ could become, if it is in the most sensitive direction, an error as large as $\epsilon/|\lambda_{\min}|$ in the solution $x$, where $\lambda_{\min}$ is the eignevalue of $A$ with the smallest modulus. On the other hand if $b$ is in the least sensitive direction, the solution $x$ could be as small as $b/|\lambda_{\max}|$. Thus a relative error in the input vector $b$ could increase by factor

$$K = \frac{|\lambda_{\max}|}{|\lambda_{\min}|}$$

in the solution vector $x$. This ratio of the largest to smallest eigenvalue of $A$ is called the 'condition number' of matrix $A$. It can be quite large. This increase in the error is unavoidable.

Theoretically Gaussian elimination, $LU$ decomposition, can create much larger errors. Theoretically, rounding errors can double as each row is processed, leading to an amplification of $2^n$. Fortunately, such systematic accumulation is rarely seen, and instead there is much cancellation of random errors. Also dividing by small pivots $a_{kk}$ can amplify the error, and this would be minimised by pivoting.

## 8.3 *QR* decomposition

In the $QR$ decomposition of a matrix $A$, $Q$ is an orthogonal matrix and $R$ is an upper triangular matrix. This decomposition is an alternative to $LU$ decomposition for solving linear systems of equations. It is thought that the rotation and or reflection by $Q$ does not stretch and increase errors like $LU$ decomposition. An orthogonal matrix has its columns orthogonal to one another, $QQ^T = I$.

This says that the inverse of the orthogonal matrix is simply its transpose, $Q^{-1} = Q^T$, so that there is no cost in calculation the inverse. Hence one can solve the linear system of equations $Ax = b$ back substitution

$$Rx = Q^T b.$$

The $QR$ decomposition is also used in the calculation eigenvalues. We note here that, because the determinant of an orthogonal matrix $Q$ is $\pm 1$ and the determinant of an upper triangular matrix is just the product of its diagonal elements, then

$$\det A = \pm \Pi_i r_{ii}.$$

The orthogonal matrix $Q$ is not unique. This section gives three different methods for finding different orthogonal matrices $Q$, by Gram–Schmidt orthogonalisation, by Givens rotations and by Householder reflections.

### 8.3.1  *QR* by Gram–Schmidt

To make a $QR$ decomposition of $n \times n$ matrix $A$, one first writes the columns of $A$ as $n$ vectors $\mathbf{a}_1, \mathbf{a}_2, \ldots, \mathbf{a}_n$, each of length $n$. The standard Gram–Schmidt orthogonalisation process is then started by dividing the first column $\mathbf{a}_1$ by its length in order to produce the unit vector $\mathbf{q}_1$ which will become the first column of the orthogonal matrix $Q$. The second and subsequent columns of $Q$ are formed from the second and subsequent columns of $A$ by first projecting out components in the directions of previous columns of $Q$ and then normalisations, i.e.

$$
\begin{aligned}
\mathbf{q}_1' &= \mathbf{a}_1 & \mathbf{q}_1 &= \mathbf{q}_1'/|\mathbf{q}_1'|, \\
\mathbf{q}_2' &= \mathbf{a}_2 \;-(\mathbf{a}_2 \cdot \mathbf{q}_1)\mathbf{q}_1 & \mathbf{q}_2 &= \mathbf{q}_2'/|\mathbf{q}_2'|, \\
\mathbf{q}_3' &= \mathbf{a}_3 \;-(\mathbf{a}_3 \cdot \mathbf{q}_1)\mathbf{q}_1 \;-(\mathbf{a}_3 \cdot \mathbf{q}_2)\mathbf{q}_2 & \mathbf{q}_3 &= \mathbf{q}_3'/|\mathbf{q}_3'|, \\
\vdots &
\end{aligned}
$$

The orthogonal matrix $Q$ has as its columns the assembly of the orthogonal unit vectors $\mathbf{q}_1, \mathbf{q}_2, \ldots, \mathbf{q}_n$. The upper triangular matrix $R$ is defined by the projections

$$r_{ii} = |\mathbf{q}_i'|, \quad \text{and} \quad r_{ij} = \mathbf{a}_j \cdot \mathbf{q}_i, \quad i < j.$$

Then the orthogonalisation process above can be rewritten as

$$\mathbf{a}_j = \sum_{i=1}^{j} \mathbf{q}_i r_{ij} \quad \text{i.e.} \quad A = QR.$$

The algorithm as described is numerically unstable. In large matrices it is probable that two of the original columns of $A$ are nearly parallel. The orthogonalisation of the second of the two will then have some coefficients very nearly equal to those of the first, and this is where accuracy is lost. This can be avoided by projecting $\mathbf{q}_i$ out of remaining columns $\mathbf{a}_j$ $j > i$ when $\mathbf{q}_i$ is first formed, rather projecting out all the previous $\mathbf{q}_i$ $i < j$ when $\mathbf{a}_j$ is considered, i.e.

For $i = 1 \to n$:
$$r_{ii} = |\mathbf{a}_i|, \quad \mathbf{q}_i = \mathbf{a}_1 / r_{ii},$$
$$\text{for } j = i + 1 \to n: \quad r_{ij} = \mathbf{q}_i \cdot \mathbf{a}_j, \quad \mathbf{a}_j \leftarrow \mathbf{a}_j - r_{ij} \mathbf{q}_i.$$

### 8.3.2 *QR* by Givens rotations

For this $QR$ decomposition, the orthogonal matrix $Q$ is generated as the product of $\frac{1}{2}n(n-1)$ simple rotations. The simple rotations are of the form

$$
G_{ij} = \begin{pmatrix}
1 & & & & & & & \\
 & 1 & & & & & & \\
 & & \cos\theta & & & \sin\theta & & \\
 & & & 1 & & & & \\
 & & & & 1 & & & \\
 & & -\sin\theta & & & \cos\theta & & \\
 & & & & & & 1 & \\
 & & & & & & & 1
\end{pmatrix}
$$

Applying this to $A$ in $G_{ij}A$ alters rows $i$ and $j$

$$a_{ik} \leftarrow a_{ik}\cos\theta + a_{jk}\sin\theta, \quad \text{and} \quad a_{jk} \leftarrow -a_{ik}\sin\theta + a_{jk}\cos\theta \quad \text{for} \quad k = 1 \to n.$$

One chooses $\theta$ to zero the off-diagonal element below the diagonal $a_{ji}$ with $j > i$, i.e. chooses $\tan\theta = a_{ji}/a_{ii}$. Note that one does not need to calculate the angle $\theta$, one just needs $\cos\theta$ and $\sin\theta$, which can readily be found directly from $\tan\theta$. One needs a strategy of eliminating the lower off-diagonal elements without undoing previous zeros. One strategy is to start at the top of the first column with $G_{12}$ to zero $a_{21}$, then work down the first column with $G_{13}, G_{14}, \ldots, G_{1n}$, then to tackle the second column with $G_{23}, G_{24}, \ldots, G_{2n}$, and then continue in order through the columns. In sparse matrices, many of these rotations would be unnecessary. Having eliminated all the lower off-diagonal elements one has the upper triangular matrix $R$. The orthogonal matrix $Q$ is just the product of the inverses of the rotations in the order which they were produced

$$Q = G_{n-1\,n}^T \ldots G_{2n}^T \ldots G_{24}^T G_{23}^T G_{1n}^T \ldots G_{13}^T G_{12}^T.$$

The $QR$ decomposition by Givens rotations is numerically stable. It is a little slower than Gram–Schmidt orthogonalisation, with $2n^3$ operators compared to $\frac{1}{2}n^3$. However as one rotation only operates on two rows, it is possible to parallelise the calculation by operating on several disjoint pairs at the same time.

### 8.3.3 $QR$ by Householder reflections

For this $QR$ decomposition, the orthogonal matrix $Q$ is generated as the product of $n$ reflections. The reflection in the plane perpendicular to the vector $\mathbf{h}$ is

$$H = \left(I - 2\frac{\mathbf{h}\mathbf{h}^T}{\mathbf{h}\cdot\mathbf{h}}\right),$$

so

$$Hx = \begin{cases} \mathbf{x} & \text{if } \mathbf{x} \text{ is orthogonal to } \mathbf{h}, \\ -\mathbf{x} & \text{if } \mathbf{x} \text{ is parallel to } \mathbf{h}. \end{cases}$$

Note that reflecting twice through the same plane leaves everything unchanged, so that the inverse of this particular reflection is itself, $H^{-1} = H$.

The first reflection $H_1$ is constructed out of the first column $\mathbf{a}_1$ of $A$, as follows. Take

$$\mathbf{h}_1 = \mathbf{a}_1 + (\alpha_1, 0, \ldots, 0)^T, \quad \text{with} \quad \alpha_1 = |\mathbf{a}_1|\text{sign}(a_{11}).$$

Then we have

$$\mathbf{h}_1 \cdot \mathbf{a}_1 = |\mathbf{a}_1|^2 + |a_{11}||\mathbf{a}_1|, \quad \text{and} \quad \mathbf{h}_1 \cdot \mathbf{h}_1 = 2\mathbf{h}_1 \cdot \mathbf{a}_1.$$

The use of the sign($a_{11}$) in $\alpha_1$ is to avoid the possibility of $\mathbf{h}_1 \cdot \mathbf{a}_1 = 0$. Constructing the first reflection out of this $\mathbf{h}_1$, we have that its action on the first column of matrix $A$ is to zero all the lower off-diagonal elements

$$H_1\mathbf{a}_1 = (-\alpha_1, 0, \ldots, 0)^T.$$

The first reflection $H_1$ is then applied to the remaining column vectors of $A$. Note that applying the reflection to a vector $\mathbf{a}_k$ is only $2n$ operations rather than $n^2$ operations if one first forms the product $\mathbf{h}_1 \cdot \mathbf{a}_k$. Having zeroed the off-diagonal elements of the first column, one now omits the first column and first row and repeats the construction for the smaller $(n-1)\times(n-1)$ subsystem to form $\mathbf{h}_2$ (with $(\mathbf{h}_2)_1 = 0$) and $H_2$. This continues with ever smaller subsystems done to form the last $H_{n-1}$. The orthogonal matrix $Q$ is then just the product of these reflections in the order that they were produced

$$Q = H_{n-1}\ldots H_2H_1.$$

The $QR$ decomposition by Householder reflection requires $\frac{4}{3}n^3$ operations, so is faster than by Givens rotations, but not quite as accurate.

## 8.4 Sparse matrices

For large sparse systems such as those that come from partial differential equations, the Gaussian elimination and $QR$ algorithms are expensive. Instead of such 'direct' methods one uses iterative methods to obtain an approximate answer, such as SOR for solving the Poisson problem in Part I.

If most of the elements of $A$ are zero, then it is better to store just the non-zero elements in a 'packed' form. Evaluating the operation of a sparse matrix on a vector, $Ax$, can be much cheaper if the packed form is used. For example, the Poisson problem on an $N \times N$ grid involves an $N^2 \times N^2$ unpacked matrix with only $5N^2$ nonzero elements, so that evaluating $Ax$ is only $5N^2$ operations for the packed form compared with $N^4$ operations for the unpacked form.

There are several iterative methods suitable for sparse matrices. The important conjugate gradient method is the subject of the next section. Some methods, such as SOR, come from splitting the matrix into two additive parts, $A = B + C$ where $B$ has a simple and sparse inverse. Then the iterative method

$$x_{n+1} = B^{-1}(b - Cx_n),$$

converges if $|B^{-1}C| < 1$. Before starting such an interative process, it can be useful to *precondition* the system of equations by multiplying by a preconditioning matrix $P$, as in

$$PAx = Pb.$$

The matrix $P$ should leave the system still sparse.

## 8.5 Conjugate gradients

The conjugate gradients method is in fact a direct method, which obtains the exact answer after $n$ steps for an $n \times n$ matrix, although in practice it converges to a very good approximation well before completing all $n$ steps.

The conjugate gradients method is only for real, symmetric, positive definite matrices $A$.

To solve the linear system of equations $Ax = b$, one seeks the minimum of the quadratic

$$f(x) = \tfrac{1}{2}(Ax - b)^T A^{-1}(Ax - b) = \tfrac{1}{2}x^T Ax - x^T b + \tfrac{1}{2}b^T Ab.$$

The gradient of $f(x)$ is

$$\nabla f = Ax - b,$$

and this vanishes at the minimum.

From an approximation $x_k$, one looks for the minimum in a direction $u$. At $x = x_k + \alpha u$,

$$f(x_k + \alpha u) = f(x_k) + \alpha u \cdot \nabla f_k + \tfrac{1}{2}\alpha^2 u^T A u.$$

As a function of $\alpha$, the minimum occurs at $\alpha = -u \cdot \nabla f_k / u^T A u$.

The question now arises how to choose the direction $u$ to search for a minimum. It turns out that choosing the steepest descent $u = \nabla f$ is not a good idea. If one always chooses the steepest descent, one tends to rattle from side to side across a steep valley with no movement along the valley floor. Instead, one needs a new direction $v$ which does not undo the previous minimisation in the direction $u$. So consider moving from the approximation $x_k$ first in the direction $u$ and then in direction $v$

$$f(x_k + \alpha u + \beta v) = f(x_k) + \alpha u \cdot \nabla f_k + \tfrac{1}{2}\alpha^2 u^T A u$$
$$+ \alpha\beta u^T A v + \beta v \cdot \nabla f_k + \tfrac{1}{2}\beta^2 v^T A v.$$

In order that the second move in the direction $v$ does not interfere with the previous move in direction $u$, one needs the cross term to vanish, i.e.

$$u^T A v = 0.$$

If the two directions satisfy this, they are called conjugate directions. One takes for $v$ the gradient $\nabla f$ after removing that component in the direction $u$ so that the two directions are conjugate.

The conjugate gradient algorithm starts with an approximation $x_0$ and an initial direction which might as well be the gradient $u_0 = \nabla f(x_0)$. Denote the residual after $k$ iterations by $r_k = Ax_k - b = \nabla f_k$. The iteration to the next step proceeds as follows.

   i. move $x_{k+1} = x_k + \alpha u_k$ with minimising $\alpha = -u_k^T r_k / u_k^T A u_k$,
   ii. update the residual $r_{k+1} = r_k + \alpha A u_k$,
   iii. new conjugate direction $u_{k+1} = r_{k+1} + \beta u_k$ with $\beta = -r_{k+1}^T A u_k / u_k^T A u_k$.

Note that only one matrix evaluation $A u_k$ is made in each iteration, and this is good for sparse matrices. One can show that the new direction $u_{k+1}$ is automatically conjugate all previous directions $u_i$, $i = 1, 2, \ldots, k$, so that previous directions do not have to be remembered. One can also show that $\alpha = r_k^T r_k / u_k^T A u_k$ and $\beta = r_{k+1}^T r_{k+1} / r_k^T r_k$.

For nonsymmetric matrices there is the GMRES (General Minimal RESidual) method, which tries to minimise $(Ax-b)^T(Ax-b)$, i.e. seeks a least-square approximation for $x$. The matrix $A^TA$ has a condition number the square of the condition number of $A$, so the calculations are more sensitive to errors. The method works in the $k$-truncated *Krylov* space spanned by $b, Ab, A^2b, \ldots, A^kb$. These directions become more and more parallel, so it is a good idea to orthogonalise them as they are generated, and that requires one to remember all the previous directions. Because this becomes costly in storage and computer time, the process is sometimes stopped after a certain number of iterations $k$ and then restarted from the latest residual replacing the initial $b$.

## 8.6 Eigenproblems

As well as the standard eigenproblem of finding eigenvalues $\lambda$ and eigenvectors $e$ for a matrix $A$ in

$$Ae = \lambda e,$$

one can also be interested in the generalised eigenproblem with two matrices $A$ and $B$ in

$$Ae = \lambda Be.$$

There is no finite direct method for finding eigenvalues, except for simple $2 \times 2$, $3 \times 3$, $4 \times 4$ matrices, because there is no finite direct method of finding the roots of polynomials beyond quartics.

If the $n \times n$ matrix $A$ is real and symmetric, there exist a complete set of $n$ orthogonal eigenvectors. If the matrix is not symmetric, there are possibilities of degeneracy with duplicate eigenvalues and missing eigenvalues, all described by a nontrivial Jordan normal form. Nonsymmetric matrices also give rise to interesting nonnormal mode behaviour in fluid mechanics, as illustrated by the following simple system of coupled linear differential equations.

$$\frac{d}{dt}\begin{pmatrix} x \\ y \end{pmatrix} = \begin{pmatrix} -1 & k^2 \\ 0 & -1-k \end{pmatrix}\begin{pmatrix} x \\ y \end{pmatrix}, \quad \text{with initial conditions} \quad x(0) = 0, \ y(0) = 1.$$

This has a solution

$$x = k(e^{-t} - e^{-(1+k)t})$$

which eventually decays, but before it decays it is $k$ larger than the initial condition.

In the following sections the matrix $A$ will be real and symmetric.

## 8.7 Power iteration

This method is just to find the largest eigenvalue. One starts with a random $x_0$, and then iterates a few times

$$x_{n+1} = Ax_n = A^n x_0.$$

The vector $x_n$ becomes dominated by the eigenvector with largest eigenvalue, so that an approximate solution is

$$\lambda_{\text{approx}} = |Ax_x|/|x_n|, \qquad e_{\text{approx}} = Ax_x/|Ax_n|.$$

One can improve on this approximation by considering the matrix $(A - \lambda_{\text{approx}}I)$. This matrix has a very small eigenvalue $(\lambda_{\text{correct}} - \lambda_{\text{approx}})$. So taking the inverse, the matrix

$$\left(A - \lambda_{\text{approx}}I\right)^{-1},$$

has a very large eigenvalue $1/(\lambda_{\text{correct}} - \lambda_{\text{approx}})$. Hence the power method on this inverted matrix will converge very rapidly.

Other eigenvalues can be found using $\mu$-shifts in $(A - \mu I)^{-1}$.

## 8.8 Jacobi

This method should only be used on small matrices.

Find maximum off-diagonal, say $a_{ij}$. Make a Givens rotation $G(i, j, \theta)$ to

$$A' = GAG^T.$$

The new matrix $A'$ has the same eigenvalues as $A$. The matrix $A$ being symmetric makes the new matrix $A'$ also symmetric. The angle $\theta$ is chosen to zero $a'_{ij}$, and zero $a'_{ji}$ by the symmetry. This rotation does fill in some previous zeros, but the sum of the squares of the off-diagonals decreases by $2a_{ij}^2$. Hence repeated application converges to a diagonal form, in which the diagonal elements are the eigenvalues.

## 8.9 Main method

The standard method is made up of two steps. The first step is to make an orthogonal transformation to reduce the matrix $A$ to an upper Hessenberg form $H$, $H = Q^T A Q$. The orthogonal transformation means that $H$ has the same eigenvalues as $A$. An upper Hessenberg matrix has nonzero elements above the diagonal, on the diagonal and one below diagonal. There are several alternative methods for the second step of extracting the eigenvalues out of the Hessenberg form.

The first step of producing the upper Hessenberg matrix was introduced by Arnoldi, and is a modified Gram–Schmidt orthogonalisation of the Krylov space

$$\mathbf{q}_1, A\mathbf{q}_1, A^2\mathbf{q}_1, \ldots, A^{(n-1)}\mathbf{q}_1,$$

the modified version which is numerically stable.

One starts with an unit vector $\mathbf{q}_1$. Then

For $k = 1 \rightarrow n - 1$:
  $\mathbf{v} = A\mathbf{q}_k$,
  for $j = 1 \rightarrow k$:
    $H_{jk} = \mathbf{q}_j \cdot \mathbf{v}$,
    $\mathbf{v} \leftarrow \mathbf{v} - H_{jk}\mathbf{q}_j$,
  $H_{k+1\,k} = |\mathbf{v}|$,
  $\mathbf{q}_{k+1} = \mathbf{v}/H_{k+1\,k}$.

Hence the original $\mathbf{v} = A\mathbf{q}_k$ is expanded by the orthogonalisation processes

$$\mathbf{v} = H_{k+1\,k}\mathbf{q}_{k+1} + H_{kk}\mathbf{q}_k + \cdots + H_{1k}\mathbf{q}_1.$$

This can be written as

$$A(\mathbf{q}_1, \mathbf{q}_2, \ldots, \mathbf{q}_n) = (\mathbf{q}_1, \mathbf{q}_2, \ldots, \mathbf{q}_n)\,H,$$

or

$$AQ = QH, \quad \text{so} \quad H = Q^T A Q.$$

The above algorithm can be used for a general matrix $A$, and has a cost of $O(n^3)$ operations. If the matrix $A$ is symmetric, then the upper Hessenberg matrix $H$ will also be symmetric, and so it will be tridiagonal. Knowing that the result will be tridiagonal, one can skip the null calculations, reducing the cost to $O(n^2)$ operations. The method is then associated with the name of Lanczos.

There are several alternative methods for the second step of extracting the eigenvalues from the upper Hessenberg matrix $H$, which has the same eigenvalues as the original matrix $A$.

  i. QR decomposition.
    Find a $QR$ decomposition of $H$ and then define $H' = RQ = Q^T HA$.
    $H'$ has the same eigenvalues and remains Hessenberg/triangular,

the off-diagonals are reduced by $\lambda_i/\lambda_j$ and so repeated applications converge to a diagonal matrix.

ii. The Power method, which is rapid when $H$ is tridiagonal.

iii. Root solve $\det(A - \lambda I) = 0$, which is normally expensive and inaccurate but not so for tridiagonal matrices.

BUT USE PACKAGES, and do not code up any of the algorithms described in this chapter.

# Further reading

*Numerical linear algebra* by H. Wendland published by Cambridge University Press 2018.

# PART III

---

## SPECIAL TOPICS

# 9

# Software packages and FREEFEM++

For simple straightforward problems where one has a good idea of what the answer is, there is much to be said for using a pre-existing software package to find the exact answer. Many research problems are however not straightforward and one may have little idea of what the answer is. In these cases, it is certainly better to write one's own tailor-made code which can treat carefully the tricky points.

There are many good software packages for CFD, some free. Propriety packages include COMSOL and ANSYS FLUENT. Free packages include GERRIS a volume of fluid code, OPENFOAM a finite volume code and FREEFEM++ a finite element code. I do not have a favourite: over the years, I have given lecture demonstrations of several.

Many software packages are easy to use, despite coming with thick manuals. The remainder of this chapter is a quick introduction of one particular package, FREEFEM++. My aim is to show that packages are not difficult to use to produce useful results.

FREEFEM++ was pioneered in 1987 by Olivier Pironneau in the Laboratoire Jacques-Louis Lions in Paris, and is currently managed by Frédéric Hecht. The finite element package for 2D and 3D includes a simple but versatile mesh generator, around 40 types of finite elements, visualisation of results, MPI parallelisation capabilities, various linear solvers including sparse solvers and more. There is a nice Integrated Development Environment FREEFEM++-cs, which presents three windows, one for the code, one for figures of the results and one for the run-time commentary. There are versions for Linux, Windoze and MacOS, all of which are free and can be downloaded from the web. An online version FREEFEM++-js is now available. There is a 400-page manual and a good tutorial through worked examples.

## 9.1 Poisson problem

We consider the simple Poisson problem in a circle

$$\nabla^2 \phi = \rho \quad \text{in } r \leq 1, \quad \text{with } \phi = 0 \text{ on } r = 1.$$

We will make the calculation for the particular case of $\rho = -1.0$.

The finite element method was presented in Chapter 5. For the Poisson problem, there is a variational statement, which leads to a *weak formulation*, see §5.4,

$$\int_{r \leq 1} \nabla \phi \cdot \nabla w + \rho w = 0, \quad \text{for all } w \text{ vanishing on the boundary.}$$

This equation is how FREEFEM++ needs to be asked to solve the Poisson problem.

Below are the first three lines of the FREEFEM++ code. The first defines the boundary, a circle, and calls it 'Gamma'. Note that the domain is to the left of the curve, so it is the interior of the circle. The second line builds a mesh of triangles with 20 points around the boundary, calling the mesh 'Th'. The third plots out the mesh for inspection. The resulting plot of the mesh is shown in Figure 9.1. The lines starting with '//' are comments.

```
// define the boundary
border Gamma(t=0,2*pi) { x = cos(t); y = sin(t); };

// construct the mesh of T_h of triangles
mesh Th=buildmesh(Gamma(20));

// plot the grid
plot(Th);
```

Next the finite element space of basis functions 'Vh' is defined over the mesh 'Th', using linear variations over each triangle. Then the unknown function $\phi$ and the test functions $w$ are defined as members of that finite element space. In the third line, the function $\rho$ is also defined to be a member and to take the value $-1.0$.

```
// Finite Element space V_h of P1 (linear) elements over mesh Th
fespace Vh(Th,P1);

// unknown phi & test function w over FE space
Vh phi,w;

// set rho(x,y) = -1.0
Vh rho = -1.0;
```

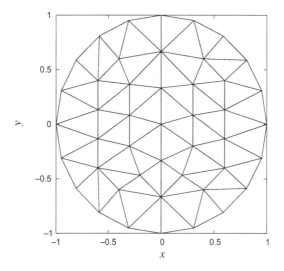

Figure 9.1 The mesh.

Now the command 'solve' is given. The problem to be solved is called 'Poisson', and is defined by the following code for the weak formulation. Note that 'Poisson' is not a predefined word. There are two two-dimensional integrals over the mesh 'Th', keeping the bilinear terms involving the unknown $\phi$ and test $w$ in one integral and the linear terms involving just $w$ in the other. The derivatives of $f$ with respect to $x$ and $y$ are dx(f) and dy(f). The final part of 'solve' specifies the boundary values. The resulting solution $\phi(x, y)$ is then plotted, and is shown in Figure 9.2.

```
// solve weak form Poisson equation with BC
solve Poisson (phi,w) = int2d(Th)(dx(phi)*dx(w) + dy(phi)*dy(w))
                      + int2d(Th)(rho*w)
                      + on(Gamma,phi=0);

// plot the result
plot(phi);
```

Figure 9.2 shows that the solution for $\phi$ is the correct general shape and magnitude. A test of accuracy is to consider the value at the centre $\phi(0, 0)$ which theoretically should be 0.25. The computed value can be obtained with two more lines of code.

```
// write to console phi(0,0), should be 0.25
x=0; y=0;
cout << phi << endl;
```

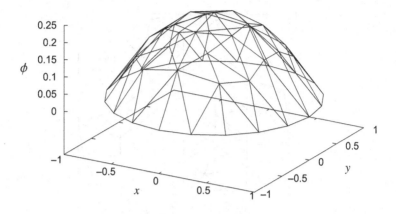

Figure 9.2 The solution of the Poisson problem.

The result for linear P1 elements and 20 points around the boundary is 0.2496. However this is unrepresentatively accurate, because with 14 points the value is 0.2545 and with 28 points 0.2484. With linear elements the value at a fixed point can be erratic, depending on how near the point is to a vertex of a triangle. Using quadratic P2 elements, the error at a fixed point decreases quadratically with the mesh size. With 20 points around the boundary, the value at the centre is 0.2455, a 2% error. One might expect the numerical result with quadratic P2 elements to be the exact analytic answer $\frac{1}{4}(1 - x^2 - y^2)$. However, the numerical boundary is not quite the circle but a polygon, and this introduces $O(\Delta x^2)$ errors.

Although of little use in this Poisson problem, one can adapt the mesh to place more points where a function $f$ is relatively large. The following six lines of code achieve this. The adapted mesh is plotted in Figure 9.3.

```
border Gamma(t=0,2*pi) { x = cos(t); y = sin(t); };
mesh Th=buildmesh(Gamma(20));
fespace Vh(Th,P1);
func f = (x-0.5)^3 + (y-0.5)^3;
Vh fh = f;
mesh Th2 = adaptmesh(Th,fh);
```

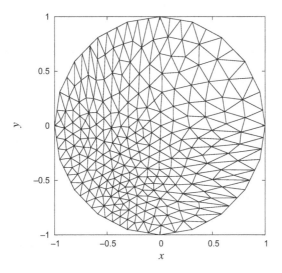

Figure 9.3 An adapted mesh placing more points where $((x - 0.5)^3 + (y - 0.5)^3)$ is large.

Finally, more complex boundaries can be constructed. The code below creates an elliptical boundary with an elliptical hole, as plotted in Figure 9.4

```
// the length of the semimajor axis and  semiminor axis
real a=2.,b=1.;
border Gamma1(t=0,2*pi)    { x = a * cos(t); y = b*sin(t); }
border Gamma2(t=0,2*pi) { x = 0.5*a * cos(t); y = 0.5*b*sin(t); }

// construction of mesh
mesh Th=buildmesh(Gamma1(30)+Gamma2(-20)); //-20 makes hole
```

## 9.2 Driven cavity

The finite element method for the Navier–Stokes equation was presented in §5.9. Here the diffusion term is treated implicitly which allows large time-steps, as we shall be interested only in the final steady state and not the time evolution. The advection term is safer and so is treated as simple first-order forward time-stepping

$$\frac{\mathbf{u}^{n+1} - \mathbf{u}^n}{\delta t} + \mathbf{u}^n \cdot \nabla \mathbf{u}^n = -\nabla p^{n+1} + \nu \nabla^2 \mathbf{u}^{n+1}.$$

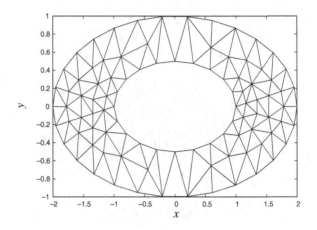

Figure 9.4 A mesh for an elliptical boundary with an elliptical hole.

The incompressibility equation is applied to the new velocity field $\mathbf{u}^{n+1}$ and to avoid spurious pressure modes an artificial pressure penalty (see §5.9.4) is introduced on the right-hand side

$$\nabla \cdot \mathbf{u}^{n+1} = -10^{-6}p^{n+1}.$$

The FREEFEM++ software requires the governing partial differential equation to be put into a weak formulation. This is obtained by multiplying the momentum equation by $\mathbf{v}$ and the incompressibility condition by $q$, adding the two, integrating over the domain and finally reducing highest derivatives by integrating by parts. Thus

$$\int \frac{1}{\delta t}\mathbf{u}^{n+1} \cdot \mathbf{v} + \nu\nabla\mathbf{u}^{n+1} : \nabla\mathbf{v} - p^{n+1}\nabla \cdot \mathbf{v} - \nabla \cdot \mathbf{u}^{n+1}q - 10^{-6}p^{n+1}q$$
$$- \frac{1}{\delta t}\mathbf{u}^{n} \cdot \mathbf{v} + \mathbf{u}^{n} \cdot \nabla\mathbf{u}^{n} \cdot \mathbf{v} = 0.$$

The streamlines can be found by solving the streamfunction-vorticity equation

$$\nabla^2\psi = -\frac{\partial v}{\partial x} + \frac{\partial u}{\partial y}.$$

Here is the FREEFEM++ code.

```
real   nu=1.0/10.0;      // i.e. Re=10
int nx = 20;             // number of points on a side
```

```
real finaltime = 3.0;
int itt = 30;              // number of time steps to stop
real dt=finaltime/itt;
int j;
int i=0;

border bottom (t=0,1.0) {x=t; y=0.0; }
border rside  (t=0,1.0) {x=1.0; y=t; }
border top    (t=0,1.0) {x=1.0-t; y=1.0; }
border lside  (t=0,1.0) {x=0.0; y=1.0-t; }

mesh Th=buildmesh(bottom(nx)+rside(nx)+top(nx)+lside(nx));
plot(Th);

fespace Xh(Th,P2);    // quadratic P2 elements for velocity
fespace Mh(Th,P1);    // linear P1 elements for pressure
Xh ux,uy,  vx,vy,  uxold,uyold;
Mh p,q;

// define Navier-Stokes problem, solve later
//    init=i to store stiffness matrix and not recompute
problem  NS ([ux,uy,p],[vx,vy,q],solver=Crout,init=i) =
    int2d(Th)( (1/dt)*(ux*vx + uy*vy)
               + nu*( dx(ux)*dx(vx) + dy(ux)*dy(vx)
               +   dx(uy)*dx(vy) + dy(uy)*dy(vy) )
               - p*(dx(vx)+ dy(vy))
               - (dx(ux) + dy(uy))*q
               - 0.000001*p*q )
  + int2d(Th)( -(1/dt)*(uxold*vx + uyold*vy)
               + (uxold*dx(uxold) + uyold*dy(uxold))*vx
               + (uxold*dx(uyold) + uyold*dy(uyold))*vy )
  + on(top,ux=sin(pi*x)*sin(pi*x),uy=0)
  + on(lside,bottom,rside,ux=0,uy=0)  ;

for (i=0;i<=itt;i++)
 { uxold=ux;       // store old time step,
   uyold=uy;
   NS;  }          // solve NS
```

```
// calculate force on top plate as intgral of shear-rate
    cout << "t = " << dt*i << " Force = " <<
        int1d(Th,top)(dy(ux))    << endl;    }

// output file of horizontal velocity on centreline x=0.5
 ofstream ofile("res.txt");
    for (j=0;j<51;j++)
    { x=0.5;y=j*0.02;
        ofile << y << "    " << ux << endl;   };

// find streamfunction
Xh psi,w;

solve streamlines(psi,w) =
        int2d(Th)( dx(psi)*dx(w) + dy(psi)*dy(w))
    +   int2d(Th)( -w*(dy(ux)-dx(uy)))
    +   on(top,lside,bottom,rside,psi=0);

plot(psi);
```

For $Re = 10$, the force is within $10^{-4}$ of its final steady value by $t = 2.0$. All the results below are for $t = 3.0$. The same steady value of the force was obtained with all time-steps $\delta t = 0.05, 0.1, 0.2$ and $0.4$. All the results below are for $\delta t = 0.1$. The pressure penalty coefficient $10^{-6}$ was reduced to $10^{-5}$ and $10^{-4}$, changing the value of the steady force by only $10^{-4}$. All the results below are for $10^{-6}$.

With 20 points along each of the sides, the value of the steady force on the top surface was found to be

$$F = 3.893.$$

This should be compared with the value of $3.8998 \pm 0.0002$ found by extrapolating to infinite resolution results from finite differences in the streamfunction-vorticity formulation of §2.10. Thus FREEFEM++ has produced a result with an error of 0.18% with 20 points along each side. With 20 points along each side, the finite difference streamfunction-vorticity algorithm produced in §2.10 a value of 3.836, i.e. a 1.8% error, while the staggered grid algorithm 3 in §3.7 produced a value of 3.851, i.e. a 1.3% error. Thus the finite element

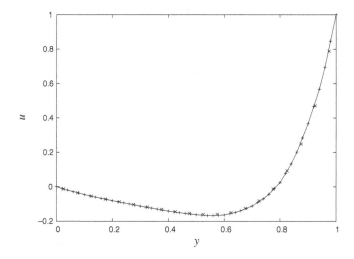

Figure 9.5 The steady-state horizontal velocity in the midsection $x = 0.5$ at $Re = 10$. The curve and + points are from FREEFEM++ with 20 points on each side, while the × points are from the finite differences in §2.10.

FREEFEM++ code is usefully more accurate at modest resolutions. As the number of points along the sides is increased, the error in the force decreases quadratically with the size of the mesh, up to 40 points on a side, and then decreases more erratically.

The horizontal velocity on the centreline $x = 0.5$ is plotted in Figure 9.5. There is a very good agreement between the new finite element results and the finite difference results in §2.10.

# 10

# Hyperbolic equations

Hyperbolic equations are very challenging to solve numerically. Their basic property is to propagate some information without changing it. Unavoidable small numerical errors made at each time-step are therefore propagated and preserved for all time. These errors tend to be systematic rather than random, so that they accumulate rather than randomly cancel. The first-order schemes of §§10.1–10.3 have numerical diffusion which erroneously spreads the propagating information, leading to a decrease in the amplitude. The second-order schemes of §§10.4–10.6 have numerical dispersion which makes information with different wavelengths propagate at different speeds, leading to spurious oscillations. Often after $O(1)$ time the propagating signal is nothing like the original signal which was to be propagated without change.

Before proceeding to tackle numerically a purely hyperbolic problem, it may be worth trying to reformulate it in terms of characteristics using a moving grid which carries the propagating information with it. For the Navier–Stokes equations, these characteristics are a Lagrangian formulation.

While pure advection is numerically challenging, advection plus diffusion is not a numerical problem so long as the spatial grid size $\Delta x$ is sufficiently small to resolve boundary layers of thickness $D/U$, where $D$ is the diffusivity and $U$ the advection speed. To resolve the fast diffusion on the fine grid scale, one also needs a small time-step $\Delta t < \Delta x^2/4D$, which combined with the first constraint gives $U\Delta t < \Delta x$. Fine-scale numerical noise rapidly decays through diffusion and so does not accumulate to dominate the answer.

Similarly advection plus reaction is not a numerical problem, so long as the spatial grid size $\Delta x$ is sufficiently small to resolve the distance moved while the reaction takes place, $U\tau$, where $\tau$ is the reaction time. Reactions tend to proceed to an end state which is not affected by small numerical errors in the propagating information.

Pure advection has the problem of accumulating numerical errors which can swamp the original information. It also has a problem of being able to propagate discontinuities and rarefaction waves. One can view the discontinuity as an unresolved diffusion boundary layer, and so the problem would not arise if the diffusion had been properly treated. Propagating discontinuities ruin higher-order schemes which rely on the existence of derivatives with finite bounded magnitudes. This chapter is split into two parts, first in §§10.1–10.6 the treatment of smooth propagating information, and second the treatment of nonsmooth, i.e. discontinuous, information. The discussion will be about schemes in which the spatial and temporal discretisations interact, either negatively or advantageously. Most of the schemes will be for spatial finite differences, because finite elements and spectral representations are more difficult to arrange for the spatial and temporal discretisations to cooperate.

For the first three-quarters of the chapter, the schemes will be presented for, and tested on, a simple one-dimensional hyperbolic equation for $u(x, t)$ satisfying

$$u_t + cu_x = 0, \quad \text{with } c > 0 \text{ and constant.}$$

Generalisations to higher dimensions and variable speeds are straightforward so long as shock waves and rarefaction waves are not formed. The smooth problem will solve the above equation subject to initial conditions

$$u(x) = \begin{cases} 4(x-1)^2(2-x)^2 & \text{in} \quad 1 \le x \le 2, \\ 0 & \text{otherwise.} \end{cases}$$

The initial conditions for the nonsmooth problem will be

$$u(x) = \begin{cases} 1 & \text{in} \quad 1 \le x \le 2, \\ 0.2 & \text{otherwise.} \end{cases}$$

The discrete finite difference solution is written as usual as

$$u_\ell^n = u(x = \ell \Delta x, t = n \Delta t).$$

## 10.1 Simplest, but unstable

For the first and simplest algorithm, one can use second-order central differencing for the spatial derivative and first-order forward time-stepping, see Figure 10.1, in order to have an explicit scheme

$$\frac{u_\ell^{n+1} - u_\ell^n}{\Delta t} = -c\frac{u_{\ell+1}^n - u_{\ell-1}^n}{2\Delta x}.$$

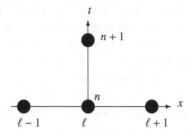

Figure 10.1 The numerical molecule for the simplest, the Lax–Friedricks and the Lax–Wendroff algorithms.

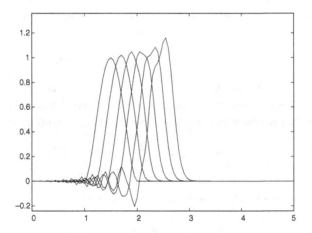

Figure 10.2 Results for the simplest algorithm at $ct = 0.0\,(0.2)\,1.0$ for $\Delta x = 0.05$ and $c\Delta t = 0.0125$. The algorithm is unstable.

In order not to propagate information further than one grid spacing in one time-step, one would expect to have to keep $c\Delta t < \Delta x$. However the algorithm is unstable for all $c\Delta t/\Delta x$. Figure 10.2 has $c\Delta t = 0.25\Delta x$, and shows that the amplitude is growing by nearly 20% by time $ct = 1.0$. There is also an erroneous negative oscillation in the tail.

*To examine the stability of the algorithms considered in this chapter*, one exploits the properties of the test equation: its linearity, constant coefficients and independence of the origin. One can then substitute a Fourier wave for the spatial variation and an exponential dependence in time, i.e. try

$$u_\ell^n = A^n e^{ik\ell\Delta x}.$$

The algorithm gives

$$A = 1 - i\mu \sin\theta, \quad \text{with} \quad \mu = \frac{c\Delta t}{\Delta x}, \quad \theta = k\Delta x.$$

It is most unfortunate that the subject of numerical analysis has adopted the symbol $\mu$ for the key parameter of the ratio of the size of the time-step to the size of the space-step, whereas the subject of fluid mechanics reserves the symbol $\mu$ for the viscosity of a fluid. This chapter is much concerned with the stability of numerical algorithms and little concerned with fluid mechanics, so we shall follow the convention of numerical analysis for this chapter.

Now the result above for $A$ has $|A| > 1$, and so the amplitude grows every time-step whatever the value of $\mu$. The most unstable wave has $\theta = k\Delta x = \frac{1}{2}\pi$, and then

$$u \sim (1 + \mu^2)^{t/2\Delta t} \quad \text{at fixed } \mu, \quad \text{or} \quad \sim e^{\frac{c^2 \Delta t}{2\Delta x^2} t} \quad \text{at fixed } \Delta x \text{ as } \Delta t \to 0.$$

In the first case of constant $\mu$, the solution blows up faster the smaller the time-step. Very unsatisfactory.

## 10.2 Lax–Friedricks, too stable

The above simplest algorithm can be easily stabilised by just replacing $u_\ell^n$ in the time derivative by that average of adjacent values $\frac{1}{2}(u_{\ell+1}^n + u_{\ell-1}^n)$. Hence

$$u_\ell^{n+1} = \frac{1}{2}\left(1 - \frac{c\Delta t}{\Delta x}\right)u_{\ell+1}^n + \frac{1}{2}\left(1 + \frac{c\Delta t}{\Delta x}\right)u_{\ell-1}^n.$$

The two brackets show that the algorithm correctly propagates the information from one grid point to the next without change if $c\Delta t = \pm\Delta x$.

To analyse the stability of the algorithm, one makes the same substitution as in the previous section,

$$u_\ell^n = A^n e^{ik\ell\Delta x}.$$

The Lax–Friedricks algorithm gives

$$A = \cos\theta - i\mu \sin\theta,$$

and so

$$|A| < 1 \quad \text{for all } \theta \text{ if } \quad \mu = \frac{c\Delta t}{\Delta x} < 1.$$

(a)                                                    (b)

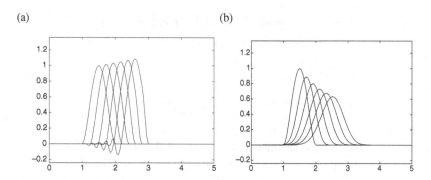

Figure 10.3 Results for the Lax–Friedricks algorithm at $ct = 0.0(0.2)1.0$ for $\Delta x = 0.05$ and (a) unstable $\mu = c\Delta t/\Delta x = 1.1$ and (b) stable $\mu = 0.5$.

This is known as the CFL (Courant–Friedricks–Lewy) condition of ensuring that information propagates less than one grid block $\Delta x$ in one time-step of $\Delta t$.

Figure 10.3 gives the results for the Lax–Friedricks algorithm for the unstable case $\mu = 1.1$ and the stable case $\mu = 0.5$. Both cases are unsatisfactory. While the solution for $\mu = 0.5$ is stable, the wave spreads and the amplitude has decreases to 60% of the initial amplitude by $ct = 1.0$. The only way to preserve correctly the amplitude is to have $A = 1$, i.e. to have $\mu = 1.0$, in which case information is passed from one grid point to the adjacent point each time-step. While this is possible for the linear problem with constant coefficients in one dimension, it is not possible in any generalisation.

To help understand the spreading wave, it is instructive to consider the behaviour in a *long-wave approximation*. A Taylor series expansion is made about about $x = \ell\Delta x$ and $t = n\Delta t$ for the terms in the algorithm, e.g.

$$u_{\ell+1}^n = u_\ell^n + \Delta x \, u_x{}_\ell^n + \tfrac{1}{2}\Delta x^2 \, u_{xx}{}_\ell^n + \cdots,$$
$$u_\ell^{n+1} = u_\ell^n + \Delta t \, u_t{}_\ell^n + \tfrac{1}{2}\Delta t^2 \, u_{tt}{}_\ell^n + \cdots.$$

Successive terms are smaller by the slow variation of $u(x, t)$ in the long-wave limit. The expansions have been continued to include the first error terms. Substituting the expansions into the Lax–Friedricks algorithm and using the Lax trick that $u_{tt} = c^2 u_{xx}$, we obtain the equation governing the numerical solution of the algorithm

$$u_t = -cu_x + \tfrac{1}{2}(1 - \mu^2)\frac{\Delta x^2}{\Delta t} u_{xx}.$$

The two error terms have both added *numerical diffusion* which totally changes the behaviour of the original hyperbolic equation. For the stable case of $\mu < 1$,

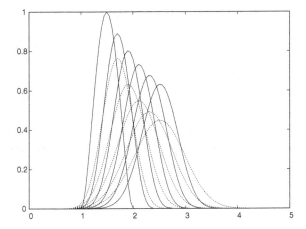

Figure 10.4 Results for Lax–Friedricks algorithm at $ct = 0.0\,(0.2)\,1.0$ for $\Delta x = 0.05$. The continuous curves are for $c\Delta t = 0.025$, while the faster spreading dashed curves are for $c\Delta t = 0.0125$.

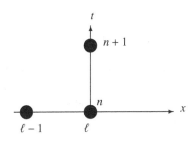

Figure 10.5 The numerical molecule for the upwinding algorithm.

this diffusion spreads the wave and thus decreases its amplitude. Note that the diffusivity increases, i.e. the wave spreads faster, as the time-step is decreased. This unsatisfactory behaviour of worse results as $\Delta t \to 0$ is illustrated in Figure 10.4, which shows faster spreading for $c\Delta t = 0.0125$ compared with $c\Delta t = 0.025$.

## 10.3 Upwinding

The first two algorithms have violated the key property of hyperbolic equations to propagate information in one direction, so-called downwind. For this chapter's simple one-dimensional test equation with speed $c$ positive, information is supposed to propagate only from $x = (\ell - 1)\Delta x$ to $x = \ell\Delta x$. But in estimating the spatial derivative, both algorithms use the value of $u$ at the downwind position $x = (\ell + 1)\Delta x$. The upwinding algorithm of this section avoids this error by estimating the spatial derivative with a one-sided derivative, the one side being upwind, see Figure 10.5,

$$\frac{u_\ell^{n+1} - u_\ell^n}{\Delta t} = -c\frac{u_\ell^n - u_{\ell-1}^n}{\Delta x}.$$

This algorithm is first-order accurate in space and time.

Applying the stability analysis of the two proceeding sections, one finds

$$|A|^2 = 1 - 4\mu(1 - \mu)\sin^2 \tfrac{\theta}{2}.$$

The algorithm is therefore stable, i.e. $|A| < 1$ for all $\theta$ if $\mu < 1$, i.e. the CFL condition is satisfied.

Applying the long-wave analysis of the first error terms, similar to that in the preceding section, one finds that the numerical solution is again governed by a diffusion equation

$$u_t = -cu_x + \tfrac{1}{2}(1 - \mu)c\Delta x\, u_{xx}.$$

Unlike the numerical diffusivity of the Lax–Friedricks algorithm which increases unboundedly as $\Delta t \to 0$, the numerical diffusivity of the upwinding algorithm is independent of $\Delta t$. Figure 10.6 gives the results for the upwinding algorithm, showing the effects of the numerical diffusion.

## 10.4 Crank–Nicolson

The Crank–Nicolson algorithm forgets the issue of propagating information only downwind and instead is second-order accurate in both space and time, see Figure 10.7,

$$\frac{u_\ell^{n+1} - u_\ell^n}{\Delta t} = -\frac{c\Delta t}{4\Delta x}\left(u_{\ell+1}^{n+1} - u_{\ell-1}^{n+1} + u_{\ell+1}^n - u_{\ell-1}^n\right).$$

The algorithm is implicit, requiring a tridiagonal system of equations to be solved for $u^{n+1}$.

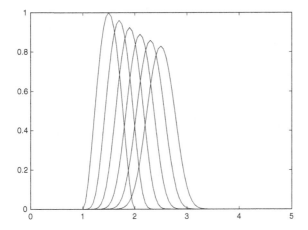

Figure 10.6 Results for upwinding algorithm at $ct = 0.0\,(0.2)\,1.0$ for $\Delta x = 0.05$ and $c\Delta t = 0.025$.

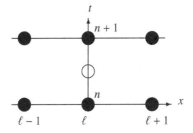

Figure 10.7 The numerical molecule for the Crank–Nicolson algorithm. The algorithm uses central differencing about the midpoint $(\ell, n + \frac{1}{2})$.

The standard stability analysis gives

$$A = \frac{1 - \frac{1}{2} i\mu \sin\theta}{1 + \frac{1}{2} i\mu \sin\theta}.$$

Hence $|A| = 1$ for all $\mu$, i.e. the algorithm is stable for all $\mu$. Moreover because $|A| = 1$ the amplitude does not decay in time. The phase of $A$ determines the numerical speed of propagation and this can be significantly different from the correct analytic speed if $\mu$ is large. Figure 10.8 shows the smooth initial conditions propagating with a constant amplitude. The Crank–Nicolson algorithm is the first algorithm to be neither unstable nor seriously damped.

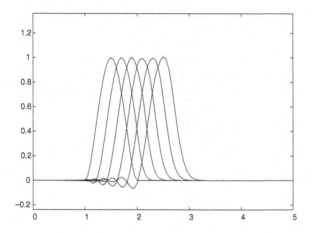

Figure 10.8 Results for the Crank–Nicolson algorithm at $ct = 0.0 (0.2) 1.0$ for $\Delta x = 0.05$ and $c\Delta t = 0.025$.

The standard long-wave analysis of the first error terms finds that the numerical solution is governed by

$$u_t = -cu_x - \frac{1}{12}(2 - \mu^2)c\Delta x^2 \, u_{xxx}.$$

This equation describes the *numerical dispersion* of different wavelengths, with long waves travelling at $c$ and shorter waves travelling slower (so long as $\mu^2 < 2$). Figure 10.8 shows slow short waves being shed at the trailing edge of the propagating initial wave, producing spurious trailing oscillations.

## 10.5 Lax–Wendroff

The Lax–Wendroff algorithm starts with the upwinding algorithm and corrects the leading order error

$$\tfrac{1}{2}(1 - \mu)c\Delta x \, u_{xx},$$

by subtracting it using the approximation $u_{xx} \approx (u_{\ell+1}^n - 2u_\ell^n + u_{\ell-1}^n)/\Delta x^2$. Thus the algorithm is

$$\frac{u_\ell^{n+1} - u_\ell^n}{\Delta t} = -c \frac{u_\ell^n - u_{\ell-1}^n}{\Delta x} - \tfrac{1}{2}(1 - \mu)c\Delta x \left( \frac{u_{\ell+1}^n - 2u_\ell^n + u_{\ell-1}^n}{\Delta x^2} \right).$$

Rearranging this becomes

$$u_\ell^{n+1} = u_\ell^n - \frac{c\Delta t}{2\Delta x}\left( u_{\ell+1}^n - u_{\ell-1}^n \right) + \frac{c^2\Delta t^2}{2\Delta x^2}\left( u_{\ell+1}^n - 2u_\ell^n + u_{\ell-1}^n \right).$$

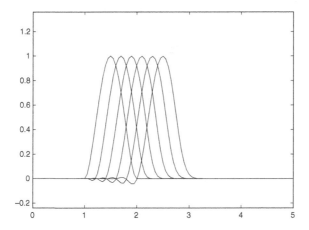

Figure 10.9 Results for the Lax–Wendroff algorithm at $ct = 0.0\,(0.2)\,1.0$ for $\Delta x = 0.05$ and $c\Delta t = 0.025$.

This algorithm is second-order accurate and explicit.

The standard stability analysis gives

$$|A|^2 = 1 - 4\mu^2(1 - \mu^2)\sin^4 \tfrac{1}{2}\theta,$$

so that the algorithm is stable if $\mu < 1$, which is the CFL condition.

The standard long-wave analysis of the first error terms finds that the numerical solution is governed by

$$u_t = -cu_x - \tfrac{1}{6}(1 - \mu^2)c\Delta x^2 u_{xxx}.$$

Again there is numerical dispersion of different wavelengths.

Figure 10.9 shows results for the Lax–Wendroff algorithm. Like the Crank–Nicolson algorithm, the amplitude remains constant and the numerical dispersion produces a trail of spurious slow short waves.

## 10.6 Angled Derivative

This algorithm respects the direction of the propagation of information like the upwinding algorithm, but achieves second-order accuracy using three time levels, $n - 1$, $n$ and $n + 1$. The idea is to use second-order accurate central differences around the midpoint $x = (\ell - \tfrac{1}{2})\Delta x$ at $t = n\Delta t$, see Figure 10.10. The time derivative is formed there by averaging the time derivatives

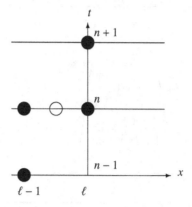

Figure 10.10 The numerical molecule for the Angled Derivative algorithm. The algorithm uses central differencing about the mid-point $(\ell - \frac{1}{2}, n)$.

at $x = (\ell - 1)\Delta x$ at $t = (n - \frac{1}{2})\Delta t$ and $x = \ell\Delta x$ at $t = (n + \frac{1}{2})\Delta t$. Thus

$$(u_t)^n_{\ell-\frac{1}{2}} = \frac{1}{2}\left(\frac{u^n_{\ell-1} - u^{n-1}_{\ell-1}}{\Delta t} + \frac{u^{n+1}_{\ell} - u^n_{\ell}}{\Delta t}\right) = -c(u_x)^n_{\ell-\frac{1}{2}} = c\frac{u^n_{\ell} - u^n_{\ell-1}}{\Delta x}.$$

Rearranging, this becomes

$$u^{n+1}_{\ell} = \left(1 - \frac{2c\Delta t}{\Delta x}\right)\left(u^n_{\ell} - u^n_{\ell-1}\right) + u^{n-1}_{\ell-1}.$$

The one difficulty of this algorithm is starting it from the initial condition at $t = 0$. One needs to generate the solution at the first time level $t = \Delta t$ either analytically by possibly several terms in a Taylor series about $t = 0$ or by using another algorithm, possibly only a first-order accurate one for a single step.

The standard stability analysis gives

$$A^2 = (1 - 2\mu)(1 - e^{-i\theta})A + e^{-i\theta}.$$

Recast as

$$\left(Ae^{i\theta/2}\right)^2 - 2i(1 - 2\mu)\sin\tfrac{1}{2}\theta\left(Ae^{i\theta/2}\right) - 1 = 0,$$

it is clear that the algorithm is stable if $\mu < 1$, the CFL condition again. The quadratic equation for $A$ means that there are two branches of solutions, one corresponding to the solution of the original partial differential equation which starts near $A = +1$ when $\theta$ is small and a spurious one which starts near $A = -1$

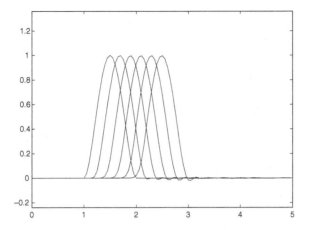

Figure 10.11 Results for the Angled Derivative algorithm at $ct = 0.0\,(0.2)\,1.0$ for $\Delta x = 0.05$ and $c\Delta t = 0.0125$.

when $\theta$ is small. The spurious mode therefore oscillates rapidly in time. However because $|A| = 1$ when stable, the amplitude of the spurious mode does not grow in time.

The standard long-wave analysis of the first error terms finds that the numerical solution is governed by

$$u_t = -cu_x + \tfrac{1}{12}(1 - \mu)(1 - 2\mu)c\Delta x^2\,u_{xxx}.$$

Note that the numerical dispersion vanishes at $\mu = \tfrac{1}{2}$. The long waves travel slower than short waves if $\mu < \tfrac{1}{2}$, and faster if $\tfrac{1}{2} < \mu < 1$.

Figure 10.11 shows the results for the Angled Derivative algorithm. As in the last two algorithms, the amplitude is preserved. Numerical dispersion for the Angled Derivative algorithm produces a trail of spurious fast short waves leading the main wave because in this case $\mu < \tfrac{1}{2}$.

Six algorithms have been presented in §§10.1–10.6. In general they are stable so long as the CFL condition $\mu = c\Delta t/\Delta x < 1$ is satisfied, which requires information not to propagate completely across a spatial grid block within one time-step. The first-order accurate algorithms §§10.1–10.3 are not particularly successful. They suffer from artificial numerical diffusion which quite quickly spreads the information, decreasing its amplitude. On the other hand the second-order accurate algorithms §§10.4–10.6 are much better. They preserve the amplitude fairly well but suffer from artificial numerical dispersion which produces spurious oscillations. These conclusions are however for

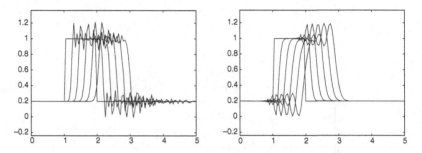

Figure 10.12 Results on the left for the Angled Derivative algorithm and on the right for the Lax–Wendroff algorithm at $ct = 0.0\,(0.2)\,1.0$ for $\Delta x = 0.05$ and $c\Delta t = 0.0125$, for nonsmooth initial conditions.

smooth initial conditions. We shall see the situation is somewhat reversed for nonsmooth initial conditions.

## 10.7 Propagation of discontinuities

The above algorithms are now applied to the nonsmooth initial conditions

$$u(x) = \begin{cases} 1 & \text{in } 1 \le x \le 2, \\ 0.2 & \text{otherwise.} \end{cases}$$

Thus there are two discontinuities to be propagated. Figure 10.12 shows the results for the second-order accurate Angled Derivative and Lax–Wendroff algorithms. While a pulse is basically propagating at the correct speed, it is developing increasing spurious oscillations at the two discontinuities which are ruining the solution. On the other hand, Figure 10.13 shows the results for the first-order upwinding algorithm. There are no spurious oscillations. The discontinuities are spreading out, but the pulse remains more or less recognisable. Clearly a new idea is needed.

To avoid the spurious oscillations, one needs to use algorithms with the property called *Total Variation Diminishing* (TVD). The total variation is the sum of the absolute value of the differences between adjacent points

$$TV(u^n) = \sum_\ell |u^n_{\ell+1} - u^n_\ell|.$$

The total variation is therefore the sum of all the differences between adjacent minima and maxima, and so independent of the numerical resolution. A TVD

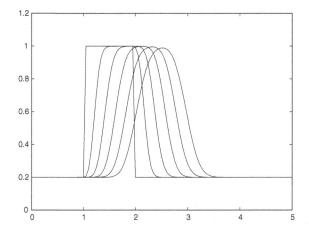

Figure 10.13 Results for the Upwinding algorithm at $ct = 0.0\,(0.2)\,1.0$ for $\Delta x = 0.05$ and $\Delta t = 0.0125$, for nonsmooth initial conditions.

algorithm has the property that the total variation does not increase in time

$$TV(u^{n+1}) \leq TV(u^n).$$

A TVD algorithm will not introduce spurious oscillations with new minima and maxima: a TVD algorithm preserves the monotonicity of a section of the solution.

## 10.8 Flux limiters

The new idea needed to treat better the propagation of discontinuities is to split an algorithm into a lower-order part, e.g. upwinding, plus a higher-order correction, e.g. the Lax–Wendroff correction. The higher-order part is then 'limited', suppressed in an oscillation and reduced if the gradient is changing rapidly. The regions where the higher-order part is suppressed are normally small, so that the first-order errors from these regions should have a small overall effect on the solution.

The hyperbolic equation is first reformulated in conservation form with a divergence of fluxes $f$

$$u_\ell^{n+1} = u_\ell^n - \frac{\Delta t}{\Delta x}\left(f_{\ell+\frac{1}{2}}^n - f_{\ell-\frac{1}{2}}^n\right).$$

For an algorithm based on upwinding plus the Lax–Wendroff correction, the fluxes are given (for $c > 0$) by

$$f^n_{\ell+\frac{1}{2}} = cu^n_\ell + \tfrac{1}{2}c(\Delta x - c\Delta t)u'_{\ell+\frac{1}{2}},$$

where $u'_{\ell+\frac{1}{2}} = \frac{u^n_{\ell+1} - u^n_\ell}{\Delta x}$ is to be limited by the upstream $\frac{u^n_\ell - u^n_{\ell-1}}{\Delta x}$.

If $c < 0$, the upstream side switches and the flux would be

$$f^n_{\ell+\frac{1}{2}} = cu^n_{\ell+1} + \tfrac{1}{2}|c|(\Delta x - |c|\Delta t)u'_{\ell+\frac{1}{2}},$$

where $u'_{\ell+\frac{1}{2}} = \frac{u^n_{\ell+1} - u^n_\ell}{\Delta x}$ is to be limited by the upstream $\frac{u^n_{\ell+2} - u^n_{\ell+1}}{\Delta x}$.

There are a number of schemes for limiting the fluxes. Let $a$ be the value to be limited by the upstream value $b$. The Minmod scheme is

$$\text{Minmod}(a, b) = \begin{cases} 0 & \text{if} \quad ab < 0, \\ a & \text{if} \quad ab > 0 \quad \text{and} \quad |a| < |b|, \\ b & \text{if} \quad ab > 0 \quad \text{and} \quad |b| < |a|. \end{cases}$$

Thus there is no high-order correction if there is an oscillation, $ab < 0$, while if the variation is monotone the smaller of the slopes is used. Taking the smaller of the slopes means that the flux is often limited, and so the spreading of the upwinding algorithm is often not corrected. Partly correcting this, the Superbee scheme is

$$\text{Superbee}(a, b) = \begin{cases} 0 & \text{if} \quad ab < 0, \\ a & \text{if} \quad ab > 0 \quad \text{and} \quad \left(|a| < \tfrac{1}{2}|b| \quad \text{or} \quad |b| < |a| < 2|b|\right), \\ b & \text{if} \quad ab > 0 \quad \text{and} \quad \left(|b| < \tfrac{1}{2}|a| \quad \text{or} \quad |a| < |b| < 2|a|\right). \end{cases}$$

Here there is again no higher-order correction if there is an oscillation, while if the variation is monotone the larger value is used so long as it is less than twice the smaller, otherwise the smaller is used. Hence the higher-order correction is applied when the slope is changing gradually. Both the Minmod and Superbee flux limiters produce TVD algorithms.

Figure 10.14 gives the results of the algorithm with the Minmod and the Superbee flux limiters. Both preserve the pulse with less spreading than the uncorrected upwinding algorithm. As expected, the Superbee gives slightly sharper fronts where the discontinuities should be.

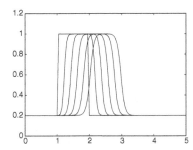 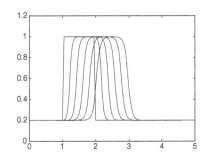

Figure 10.14 Results for the algorithm on the left with the Minmod flux limiter and on the right with the Superbee flux limiter at $ct = 0.0\,(0.2)\,1.0$ for $\Delta x = 0.05$ and $c\Delta t = 0.0125$, for nonsmooth initial conditions.

## 10.9 Nonlinear advection

So far the speed of propagation, $c$, has been constant. The generalisation to a speed which varies in space and time, $c(x, t)$, is straightforward. We now turn to the more difficult case of nonlinear advection in which the speed depends on the value being propagated, $c(u)$. Normally bigger things move faster, and $c$ is an increasing function of $u$, as in water waves. However, there are systems where smaller things move faster, $c$ is a decreasing function of $u$, for example sand dunes.

Two new problems arise with nonlinear advection, one avoidable, one not. The unavoidable problem occurs when some fast values are following some slow values. After a finite time the fast will overtake the slow, and at that instant a shockwave is formed. The discontinuity of a shockwave is challenging numerically. The avoidable other problem occurs when the initial conditions have a discontinuity with fast things starting ahead of slow things. Obviously a gap is going to open up between the fast and the slow. This gap is filled by a so-called rarefaction wave or expansion fan. This problem is avoidable, because instead one can start the mathematical idealisation of the discontinuity spread over a small region. In this small region all the values would occur between the fast and the slow, which in time would nicely fill the gap between the fast speeding ahead of the slow.

To study nonlinear advection, the propagation form of the hyperbolic equation

$$u_t + c(u)u_x = 0$$

is best reformulated in the conservation form of a divergence of fluxes $f(u)$

$$u_t + (f(u))_x = 0,$$

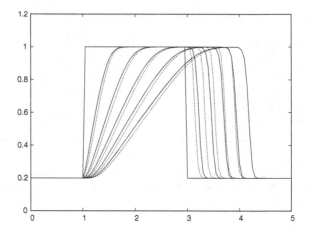

Figure 10.15 Results for nonlinear advection with flux $f(u) = \frac{1}{2}u^2$ using the upwinding algorithm at $ct = 0.0\,(0.4)\,2.0$ for $\Delta x = 0.05$ and $\Delta t = 0.0125$. The continuous curves are from the conservation formulation while the dashed curves are from the propagation formulation.

the connection being $c(u) = df/du$. Using the conservative form produces the correct speed of the shock waves, which in aeronautics means getting the correct distribution of pressure over a wing. The conservative formulation is sometimes called a *Finite Volume* or a *Volume of Fluid* approach.

While nonlinear advection is only propagating information, the different values propagating at different speeds make the waveform evolve in time. Because something definite is happening compared with a pulse propagating without change of form, it is safer to use first-order accurate algorithms, and in particular the upwinding algorithm. Small spreading from the numerical diffusion of a first-order algorithm has little effect on the naturally spreading of the rarefaction wave. Shock waves are self-sharpening, and so the effect of small numerical diffusion is only to spread the discontinuity of the shock over a couple of grid blocks.

Figure 10.15 gives the results for nonlinear advection using the upwinding algorithm. The flux function is $f(u) = \frac{1}{2}u^2$, so that the propagation form is the simple nonlinear hyperbolic equation $u_t + uu_x = 0$. The figure shows a shockwave at the front and a rarefaction wave at the rear. In between there is a propagating region of the constant $u = 1$ of the initial conditions. This region decreases in size and eventually disappears because it is propagating faster than the shockwave ahead of it.

A shockwave between values $u_L$ behind and $u_R$ ahead propagates at a speed

$$V = \frac{f(u_L) - f(u_R)}{u_L - u_R},$$

by conservation of flux in the frame moving with the shock. For the example in Figure 10.15 with $u_L = 1.0$ and $u_R = 0.2$, this expression gives the speed of the shock as $V = 0.6$. The continuous curves of the conservation form have a speed of 0.59, which is good, while the dashed curves of the propagation form have a speed of 0.46, which is unacceptably poor. There are three points within the shockwave fronts.

## 10.10 Godunov method

The Godunov method has three steps: R, E and A.

R: *Reconstruct* the solution into a simple form. Normally the simple form is piecewise constant, that is constant in each grid block. Occasionally piecewise linear might be used. Note that there will be discontinuities at the boundaries of the grid blocks.

E: The simple form is *evolved* exactly. Constant parts are advected at a constant speed. The discontinuities are propagated as shock waves or rarefaction waves. The time-step must be limited by the CFL condition to stop discontinuities propagating through more than one grid block.

A: The resulting function is *averaged* over grid blocks in preparation for step R of the next time-step.

Note that the details of the second step can be skipped. In a conservative formulation to calculate the change in one time-step of the average value in one grid block, one only needs to know the difference in the fluxes over the two ends. These fluxes can be calculated from the values of the solution on the upstream side. In the case of the piecewise constant reconstruction, this is particularly simple and leads to the upwinding algorithm in the conservative form that was discussed in the previous section.

For a general flux law $f(u)$, there is a question as to which side is upstream. Information propagates at speed $f'(u)$. When the speeds are of the same sign in adjacent grid blocks, the choice of the upstream side is straightforward. Thus for the flux $f_{\ell+\frac{1}{2}}$ between $\ell$th and $\ell + 1$th grid blocks, one would use $f(u_\ell)$ if $f'(u_\ell) > 0$ and $f'(u_{\ell+1}) > 0$, and use $f(u_{\ell+1})$ if $f'(u_\ell) < 0$ and $f'(u_{\ell+1}) < 0$. If the two speeds have different signs and are towards one another, $f'(u_\ell) > 0$ and $f'(u_{\ell+1}) < 0$, there will be a shock wave travelling at the velocity $V$ given at the end of the last section. One would then use $f(u_\ell)$ if $V > 0$ and $f(u_{\ell+1})$ if $V < 0$. If the two speeds have different signs and are away from one another, $f'(u_\ell) < 0$ and $f'(u_{\ell+1}) > 0$, there will be a rarefaction wave, and in this case one would use the flux $f(u)$ for the value $u = u_*$ which does not propagate $f'(u_*) = 0$. Thus

$$
f_{\ell+\frac{1}{2}} = \begin{cases}
f(u_\ell) & \text{if} \quad f'(u_\ell) > 0, f'(u_{\ell+1}) > 0, \\
f(u_{\ell+1}) & \text{if} \quad f'(u_\ell) < 0, f'(u_{\ell+1}) < 0, \\
f(u_\ell) & \text{if} \quad f'(u_\ell) > 0, f'(u_{\ell+1}) < 0, V > 0, \\
f(u_{\ell+1}) & \text{if} \quad f'(u_\ell) > 0, f'(u_{\ell+1}) < 0, V < 0, \\
f(u_*) & \text{if} \quad f'(u_\ell) < 0, f'(u_{\ell+1}) > 0, \quad \text{where } f'(u_*) = 0.
\end{cases}
$$

The Gudonov method is first-order in space and in time, through the averaging at each time-step. Further this averaging leads to significant numerical diffusion, typically with diffusivity $O(\Delta x^2/\Delta t)$. Higher-order methods replace the piecewise constant simple form with piecewise linear. There are several choices for the slopes to be used in the linear variation: central differencing the block-averaged values of the two adjacent blocks is associated with the name of *Fromm*, upwind slopes with *Beam-Warming* and downstream slopes with *Lax–Wendroff*. These slopes can be slope-limited by the Minmod, Superbee or similar schemes, in order to produce algorithms with the TVD property. As hinted above, one piecewise linear scheme gives the Lax–Wendroff algorithm.

This chapter has discussed the problems of a simple hyperbolic problem, advection in one space dimension of a single scalar quantity. Generalisations rapidly become messy. Advection in one space dimension of a vector quantity is required for one-dimensional gas dynamics, where there are three coupled nonlinear hyperbolic equations for density, velocity and energy. Even if the equations were not nonlinear, the three coupled equations mean that there are three combinations of quantities, *the Riemann invariants*, which propagate at different speeds, *the characteristic velocities*, normally in different directions. Hence one direction can be upstream for one combination but downstream for another. And in a nonlinear problem, the mix of components in the invariants changes from position to position. There are so-called *Riemann solvers* which handle this, such as the one associated with the name of *Roe*, but these solvers do not work in more than one space dimension.

# Further reading

*Finite volume methods for hyperbolic problems* by R. J. LeVeque published by Cambridge University Press in 2002.

# 11

## Representation of surfaces

Some fluids problems involve moving boundaries. The boundaries may have a prescribed motion or it might be that the motion is to be determined as part of the solution of the dynamics. The fluid might move with the moving boundary, as in water gravity waves, but can move across the boundary, as with imposed suction or blowing, or with phase changes with evaporation, condensation, melting or solidification. In all these cases, there is a need to describe the evolving boundary, which in general will have a nonregular shape. This chapter is about different methods of describing surfaces, not about the evolution of the surfaces.

There are two different approaches, to mark the surface or to mark the volume of fluid contained within the surface. Each approach has advantages and disadvantages. Marking the fluid is good when there is a change in topology, as when a drop breaks into two or two drops coalesce into one. Nothing special needs doing when there is such a change of topology, because the marked fluid remains marked fluid during the topological change. On the other hand with marked surfaces, the two lists of points on different surfaces need to be merged into a single list if two drops coalesce, and thought might be needed about the order of the points on the lists.

Marking the surface produces more accurate results for the curvature of the surface, which is needed for capillary forces. Marking the surface gives an easy estimate of the separation of two drops, so that disjoining pressure can be added to fluid forces. In fact one should worry that the claimed advantage of the marked volume approach of having no problem with topological changes is a serious disadvantage of excluding the extra physics of disjoining pressure during coalescence. Finally, the linear problems of Stokes flows and of potential flows can be tackled with the boundary integral method (see Chapter 12), which calculates quantities only on the surface of the fluid, so that this method works well with the marked surface approach.

We start with marked surfaces and then proceed to marked volumes.

## 11.1 Curves in two dimensions

In two dimensions, a surface containing fluid is a curve, say $\mathbf{x}(t)$ where in this chapter $t$ is not time but a parameter tracing out the curve. Let the curve have the marked points $\mathbf{x}_i$ for $i = 1, \ldots, N$. The crudest representation of the surface would be by linear segments between the marked points

$$\mathbf{x}(t) = \mathbf{x}_i(i + 1 - t) + \mathbf{x}_{i+1}(t - i) \quad \text{in } i \le t \le i + 1.$$

The problem with linear segments is that the curvature is concentrated at the data points where the tangent direction is discontinuous. If the curvature is required, as in capillary-driven phenomena, one can use three adjacent data points to form an appropriate second derivative. As the points are unlikely to be equally spaced, this second derivative will be only first-order accurate in the spatial step size. Instead of using a linear line between pairs of points, one can use quadratics between points grouped in threes. The tangential direction is, however, still discontinuous between adjacent groups of three. Therefore this generalisation is not worth exploring.

### 11.1.1 Splines

A much better solution is to use *cubic splines*. These are a series of cubics, a different cubic in each interval between pairs of data points. Each cubic passes through the two data points at its end, say $\mathbf{x}_i$ and $\mathbf{x}_{i+1}$. The two further degrees of freedom in the cubic are constrained so that the tangent direction $\dot{\mathbf{x}}$ and the curvature $\ddot{\mathbf{x}}$ are continuous with the cubics in the adjacent intervals. Thus one has a curve

$$\mathbf{x}(t) = \mathbf{x}_i(1 - \tau)^2(1 + 2\tau) + \dot{\mathbf{x}}_i(1 - \tau)^2\tau + \mathbf{x}_{i+1}\tau^2(3 - 2\tau) - \dot{\mathbf{x}}_{i+1}\tau^2(1 - \tau),$$

where $\tau = t - i$ in $i \le t \le i + 1$. This expression goes through the data points, and has a continuous first derivative, $\dot{\mathbf{x}}$. Requiring the second derivative to be continuous at $t = i$ gives

$$\dot{\mathbf{x}}_{i-1} + 4\dot{\mathbf{x}}_i + \dot{\mathbf{x}}_{i+1} = 3\mathbf{x}_{i+1} - 3\mathbf{x}_{i-1}.$$

This is a tridiagonal matrix for the unknown values of $\dot{\mathbf{x}}_i$.

From the formula for the curve $\mathbf{x}(t)$, one can calculate the

$$\text{unit tangent } \mathbf{t} = \frac{\dot{\mathbf{x}}}{|\dot{\mathbf{x}}|}, \text{ and the curvature } \kappa = \frac{\ddot{x}\dot{y} - \ddot{y}\dot{x}}{(\dot{x}^2 + \dot{y}^2)^{3/2}}.$$

The cubic spline has fourth-order accuracy with respect to the spatial step size. Therefore the curvature calculated from the cubic spline is second-order accurate.

One can also use that formula for the curve $\mathbf{x}(t)$ to calculate the arc-length and then redistribute the points to be equally spaced in arc-length. Alternatively one can redistribute the points with a weighting function in order to concentrate points near an interesting feature.

As well as cubic splines, there are lower- and higher-order splines. The odd-order splines are joined at the fitting data points, while for stability reasons the even-order splines are best joined midway between the data points. A useful cubic spline is the $B_3$-spline which is nonzero in only four adjacent intervals, in $-2 \leq x \leq 2$

$$B_3(x) = \tfrac{1}{4}\left((x+2)_+^3 - 4(x+1)_+^3 + 6x_+^3 - 4(x-1)_+^3 + (x-2)_+^3\right),$$

where $()_+$ means set to zero if the value of the bracket is negative.

## 11.2  Surfaces in three dimensions

The crudest representation of a surface with marked points $\mathbf{x}_i$ for $i = 1, \ldots, N$, would be to connect neighbouring points into triangles, i.e. triangulate the surface, and then have flat triangles for the approximation to the surface. As with the linear segments along a curve, the problem with these flat triangles is that the curvature is concentrated in the edges and vertices where the slope is discontinuous. One can think of a simple curved surface above the flat triangles, but there are problems matching the slopes with adjacent triangles, particularly at vertices. A little simpler is to use bicubics above rectangles rather than triangles, although there are then global issues of covering a closed surface such as a sphere with rectangles. The crudest option of flat triangles may be best for problems of flow calculations, although it would be terrible for other applications such as the reflection of light.

### 11.2.1  Redistributing points

During flow, the marked points of the surface can accumulate in some regions and become sparse in other regions, so that some redistribution of the points is desirable. Redistributing the points on a surface is more difficult than moving the points along a curve. One technique is to add an artificial extra tangential velocity to the points at each time-step, the velocity being chosen to keep the sides of the triangles similar in length. This can be achieved by giving the edges

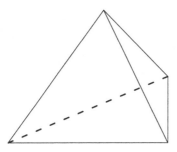

Figure 11.1 Swapping diagonals can increase the smallest angle in a pair of triangles.

a tension proportional to their length and setting the extra tangential velocity of a vertex equal to the sum of the tension forces exerted on the vertex.

If the surface is significantly deformed in the flow, the above redistribution of points can produce long thin triangles, despite trying to keep the sides at similar lengths. A second strategy to improve triangulation is to consider pairs of adjacent triangles. Swapping the diagonal from the solid diagonal line in Figure 11.1 to the dashed diagonal may produce two triangles with an increase in the smallest angle in the two triangles. Such a swap would at the same time reduce the maximum angle in the two triangles. As with adding an extra tangential velocity, this diagonal-swapping works well with time-stepping. Repeatedly applying the diagonal-swapping produces a Delaunay triangulation, one in which the circumscribing circles for each triangle contains no vertex of another triangle.

### 11.2.2 Curvature

Finding numerically the curvature of a surface marked by points is not simple. One could fit a general quadratic, $f(x, y, z) = 0$, through some marked points. A general quadratic in three dimensions has 10 degrees of freedom, so requiring fitting to 10 marked points. A little more elaborate is to set up a local plane and to express the distance from this plane to some marked points as a quadratic in two directions in the plane. The general quadratic in two dimensions has six degrees of freedom, so requiring fitting to six marked points. A problem with this approach is that numerical errors can lead to a small nonzero net capillary force on a closed surface, e.g. a net force on an isolated drop, which should not be. The following alternative approach is built on a conservation property, so that the net force on a closed surface automatically vanishes.

The capillary force on an area of the surface is equal to the net capillary

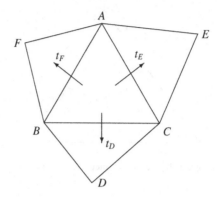

Figure 11.2 Three adjacent triangles for calculation of the capillary force on triangle ABC.

line tension exerted across the curve bounding the area. Thus instead of calculating the second derivative for the curvature, one need only calculate the first derivative for the tangent vector $\mathbf{t}$ to the surface which is normal to a curve in the surface. As long as the full surface is entirely covered by numerical subareas, the contributions from the common parts of the bounding curves cancel, ensuring zero net force on a closed surface.

### Curvature over a triangle

For the capillary force on the part of the surface represented by one triangle ABC between three marked points, one considers also the three adjacent triangles which share an edge, say *AFB*, *BDC* and *CEA* in Figure 11.2. Form the unit normals to the four triangles, say $\mathbf{n}$, $\mathbf{n}_F$, $\mathbf{n}_D$, $\mathbf{n}_E$. The tangent to the surface for the part $AB$ of the bounding curve can be estimated as $\mathbf{t}_F = \frac{1}{2}(\mathbf{n} + \mathbf{n}_F) \times \overrightarrow{AB}$. Note that this vector has the length of the side $AB$. Now the cyclic sum of the vector edges vanishes, $\overrightarrow{AB} + \overrightarrow{BC} + \overrightarrow{CA} = 0$, hence the capillary force on triangle *ABC* is

$$\frac{1}{2}\left(\mathbf{n}_F \times \overrightarrow{AB} + \mathbf{n}_D \times \overrightarrow{BC} + \mathbf{n}_E \times \overrightarrow{CA}\right).$$

From this expression, one can see that the net capillary force on a closed surface will vanish exactly, because the three triangles nearby *AFB* will have contributions $\frac{1}{2}\mathbf{n}_F \times \overrightarrow{AB}$, $\frac{1}{2}\mathbf{n}_F \times \overrightarrow{BF}$ and $\frac{1}{2}\mathbf{n}_F \times \overrightarrow{FA}$, which sum to zero.

### Curvature for a point

The above is for the capillary force associated with the face of a triangle between three marked points on the surface. For some flow calculations, it

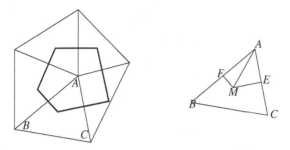

Figure 11.3 Voronoi partition about marked point $A$ using the circumcentres of the nearby triangles.

may be required to find the capillary force associated with a single marked point. For the marked point $A$, first identify all the triangles which have $A$ as a vertex. Then find the circumcentres of each of the triangles, i.e. the point where the perpendiculars through the midpoints of the sides meet. Joining the circumcentres produces the Voronoi partition for the marked point $A$, see Figure 11.3. For the capillary force exerted over the curve joining all the circumcentres, consider just one triangle, $ABC$. For the part of the curve $FM$, the tangent to the surface normal to this part of the curve is $\overrightarrow{AB}$. Now length $|FM| = \frac{1}{2}|AB|\tan\widehat{FAM}$, and angle $\widehat{FAM} = \frac{1}{2}\pi - \widehat{ACB}$. Combining with the similar contribution for the part of the curve $ME$, the contribution of triangle $ABC$ is

$$\frac{1}{2}\left(\overrightarrow{AB}\cot\widehat{ACB} + \overrightarrow{AC}\cot\widehat{ABC}\right).$$

The area of the Voronoi partition within triangle $ABC$ is

$$\frac{1}{4}\left(|AB|^2\cot\widehat{ACB} + |AC|^2\cot\widehat{ABC}\right).$$

One last idea for representing marked surfaces and calculating curvature is to try *radial basis functions*, i.e. to fit the surface as $f(\mathbf{x}) = 0$ (and obviously $f = 1$ somewhere else) using a function of the form

$$f(\mathbf{x}) = \sum_i a_i\phi(|\mathbf{x} - \mathbf{x}_i|).$$

There are many options for the basis function $\phi(r)$, with $\phi = r^3$ being the proper generalisation of cubic splines $(x-n)^3_+$ to more than one dimension. The coefficients $a_i$ are chosen so that $f(\mathbf{x}_j) = 0$ for all the marked points $\mathbf{x}_j$. This leads to a very badly conditioned linear algebra problem for the coefficients, although the fitted function is normally good. The unit normal to the surface would be $\mathbf{n} = \nabla f/|\nabla f|$ and curvature $\kappa = \nabla \cdot \mathbf{n}$. This idea works sometimes and unexpectedly fails disastrously at other times.

## 11.3  Volume of Fluid (VoF) method

This is the first of the methods which mark the volume of the fluid. Here we consider the case to two fluids, fluid 0 and fluid 1, which includes that case of fluid 0 being a vacuum. The fluids are marked by an *indicator function c*,

$$c(\mathbf{x}, t) = \begin{cases} 0 & \text{in fluid 0,} \\ 1 & \text{in fluid 1.} \end{cases}$$

The indicator function is evolved by moving it with the flow,

$$\frac{\partial c}{\partial t} + \mathbf{u} \cdot \nabla c = 0.$$

This is a hyperbolic equation which must be tackled with a good hyperbolic solver, particularly as $c$ starts as a discontinuous function. Inevitably small isolated regions are formed with $c$ negative or greater than unity. An occasional clear-out of such 'flotsam' is needed. Note one might usefully take a hint from the Level Sets of §11.5 method and choose the indicator function $c$ to vary smoothly away from the surface rather than being discontinuous.

The flow calculation is then made in the entire volume with a single 'effective' fluid, with density

$$\rho(\mathbf{x}, t) = \rho_0 + (\rho_1 - \rho_0)c(\mathbf{x}, t),$$

and similarly for the viscosity. Despite the volume of fluid 1 probably adopting an awkward moving shape, the 'effective' fluid can occupy a fixed volume with simple shape. One can then use finite differences on a Cartesian grid in order to have a fast Poisson solver. A conservative scheme on a staggered grid is used, see §4.5. The VoF method is probably the best approach for two-phase flows at intermediate Reynolds numbers.

Having marked the fluid, there is now a problem of finding the interface. Obviously the interface lies in those grid blocks where the indicator function $c$ is between 0 and 1. The interface is usually represented by a plane (in three dimensions, a line in two dimensions) cutting the grid block into two parts with volumes proportional to $c$ and $1 - c$. In the crude *Simple Linear Interface Construction* (SLIC) the cutting plane is set parallel to the axes, see Figure 11.4a. This produces a very discontinuous surface, quite unsuitable for calculation of capillary forces. Much better and commonly used is the *Piecewise Linear Interface Construction* (PLIC) in which the cutting plane is set perpendicular to $\nabla c$, see Figure 11.4b. The reconstructed surface is still discontinuous, but less so, and has the capillary force pulling in nearly the correct inclined direction.

The discontinuous indicator function $c$ is unsuitable for numerical differentiating to find $\nabla c$, so a smoothed finite difference is used. With the indicator function held at the centre of a grid block on a staggered grid, $c_{i+\frac{1}{2}\,j+\frac{1}{2}}$, the smoothed derivative in the $x$-directions is usually given by

$$\frac{\partial c}{\partial x}\bigg|_{i+\frac{1}{2}\,j+\frac{1}{2}} = \frac{1}{8\Delta x}\begin{pmatrix} -1 & 0 & 1 \\ -2 & 0 & 2 \\ -1 & 0 & 1 \end{pmatrix} c_{i+\frac{1}{2}\,j+\frac{1}{2}}.$$

The capillary force is added to the momentum balance for each grid block containing the interface by applying line tensions parallel to the cutting planes.

## 11.4 Diffuse interface method

A problem with the indicator function in the VoF method is that through numerical diffusion several grid blocks in a row can have values of $c$ gradually varying between 0 and 1. One would then have to say that the interface is in the grid block with the value nearest to $c = \frac{1}{2}$. The alternative diffuse interface method adds a term to the evolution equation for $c$ which forces the value of $c$ towards either 0 or 1, and so stops the interface becoming too diffuse. The diffuse interface method also adds a diffusion term to the evolution equation to give a specified amount of diffusion independent of uncontrolled numerical diffusion. In the diffuse interface method, the indicator function is normally called the *phase field* $\phi(\mathbf{x}, t)$ and satisfies

$$\frac{\partial \phi}{\partial t} + \mathbf{u} \cdot \nabla \phi = \epsilon_1 \nabla^2 \phi - \frac{1}{\epsilon_2} \frac{d}{d\phi}\left(\phi^2(1-\phi)^2\right).$$

This equation makes a transition region between $\phi = 0$ and 1, of thickness of $\sqrt{\epsilon_1 \epsilon_2}$, and does so on a time-scale of $\epsilon_2$.

Capillary forces can be introduced by adding the divergence of a Korteweg stress $-K\nabla\phi\nabla\phi$ to the momentum equation. This term will be active only in the thin transition region. The surface tension takes a value $\int K|\nabla\phi|^2$ integrating across the transition region.

There are problems with the diffuse interface method when applied to the flow of two continuum fluids. Similar Cahn–Hilliard equations may be appropriate for phase-change problems, but not fluid dynamics. First without any

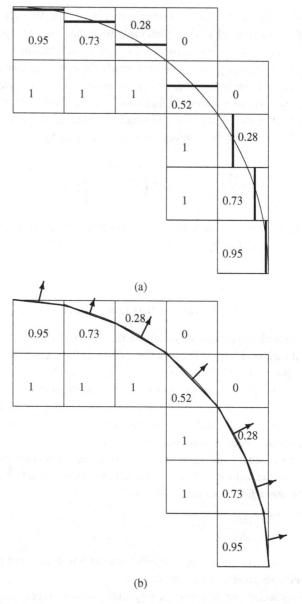

Figure 11.4 Surface reconstruction from $c(\mathbf{x})$. (a) SLIC and (b) PLIC, with vectors indicating $\nabla c$

flow the volume of a drop, a region where $\phi = 1$, can change in time. The radius $a$ of a spherical drop decreases at a rate $\dot{a} = -2\epsilon_1/a$. Second if the artificial transition region is to be thin and its dynamics fast compared with the fluid dynamics, i.e. $\epsilon_1, \epsilon_2 \ll 1$, then the numerical resolution must be much finer spatially and temporally than the flow scales. This means high numerical costs to resolve a numerical artefact which has nothing to do with the flow, and that is not smart.

## 11.5 Level sets

The really good idea is that the indicator function, normally $\phi(\mathbf{x}, t)$, starts as roughly the distance from the initial surface, taking opposite signs on the two sides of the surface. The indicator function is then moved with the flow,

$$\frac{\partial \phi}{\partial t} + \mathbf{u} \cdot \nabla \phi = 0.$$

This is a hyperbolic equation and so needs treating carefully, but the good news is that $\phi$ is a smooth continuous function and not the much more difficult discontinuous function in the first VoF method. The surface can be found at any time from

$$\phi(\mathbf{x}, t) = 0.$$

It is much easier to estimate the precise position within the interior of a grid block of the zero of $\phi$ from smooth values away from the surface, compared with the VoF indicator function which will be nearly 0 in one grid block and near 1 in the next. From time to time, the contours of $\phi$ can become crowded and no longer like distance from the surface $\phi = 0$, in which case it is good to restart the function as the distance.

### 11.5.1 Fast Marching Method

In problems where the velocity of the interface $V$ is locally determined and has a single sign, i.e. always either advances or recedes, there is a fast method for finding the shape of the evolving interface. In fluid mechanics, the velocity of the interface is determined by finding the global solution of the flow, so is not normally determined locally. Moreover in fluid mechanics, it is common for the interface to advance for some time and then recede. There are however interesting applications where $V$ is locally determined and is single signed, such as finding geodesics, illumination from a point source with obstacles, etching and deposition of microchip. Here is a quick sketch of the Fast Marching Method for those problems.

Because the interface is always advancing, the indicator function can be taken to be the time that the interface arrives at a point, i.e. arrives at $t = T(\mathbf{x})$.

This indicator function will be a smooth function of position, and will handle topological changes (interfaces merging or splitting) without difficulty. One starts with $T(\mathbf{x}) = 0$ known only on the initial interface.

- Working within a narrow band of points next to the current interface, one solves the eikonal equation

$$|\nabla T| = \frac{1}{V}.$$

- One identifies the next point to arrive on the interface.
- One can then update the list of points immediately adjacent to the interface and update their likely times of arrival on the interface.
- This list is sorted by order of time of arrival, initially by a 'heap sort', thereafter by adjusting the presorted list.

Note that the eikonal equation has problems if the speed of the interface is zero. The equation is solved in two dimensions as

$$\left( \max\left\{ \frac{T_{ij} - T_{i-1\,j}}{\Delta x}, \frac{T_{ij} - T_{i+1\,j}}{\Delta x}, 0 \right\} \right)^2$$
$$+ \left( \max\left\{ \frac{T_{ij} - T_{i\,j-1}}{\Delta x}, \frac{T_{ij} - T_{i\,j+1}}{\Delta x}, 0 \right\} \right)^2 = \frac{1}{V^2},$$

with the obvious generalisation to three dimensions.

# Further reading

*Diffuse-interface methods in fluid mechanics* by D. M. Anderson, G. B. McFadden and A. A. Wheeler in Annu. Rev. Fluid Mech. (1998) **30**, 139–165.

*Direct numerical simulation of free-surface flow* by R. Scardovelli and S. Zaleski in Annu. Rev. Fluid Mech. (1999) **31**, 567–603.

*Level set method for fluid interfaces* by J. A. Sethian and P. Smeretic in Annu. Rev. Fluid Mech. (2003) **35**, 341–372.

# 12

---

# Boundary integral method

The boundary integral method is also called the boundary element method and the panel method. It is used in linear problems with known simple Greens functions, e.g. potential flows and Stokes flows. It is essentially a way to tackle a Poisson problem. It is good for complex geometries, such as the flow outside several deforming drops. It is very good for free surface problems in which one need only find the velocity on the deforming surface.

## 12.1 Integral equation for Laplace equation

Consider the Laplace problem for $\phi(\mathbf{x})$,

$$\nabla^2 \phi = 0 \quad \text{in the volume } V,$$

$$\phi \quad \text{or} \quad \frac{\partial \phi}{\partial n} \quad \text{given on the surface } S,$$

where $\mathbf{n}$ is the unit normal to the surface out of the volume.

We need a Greens function $G(\mathbf{x}; \boldsymbol{\xi})$ satisfying

$$\nabla_x^2 G = \delta(\mathbf{x} - \boldsymbol{\xi}) \quad \text{for } \mathbf{x} \text{ in } V.$$

Here the subscript on $\nabla_x$ means differentiate with respect to $x$, while in the equation $\boldsymbol{\xi}$ is viewed as a fixed parameter. The Greens function need not satisfy any boundary condition on the surface $S$.

The Greens identity, really the divergence theorem, gives

$$\int_S \left( \phi \frac{\partial G}{\partial n} - \frac{\partial \phi}{\partial n} G \right) dS(\mathbf{x}) = \int_V \left( \phi \nabla^2 G - \nabla^2 \phi \, G \right) dV(\mathbf{x})$$

$$= \int_V \phi(\mathbf{x}) \delta(\mathbf{x} - \boldsymbol{\xi}) \, dV(\mathbf{x})$$

$$= \phi(\boldsymbol{\xi}) \times \begin{cases} 0 & \boldsymbol{\xi} \text{ outside } V, \\ 1 & \boldsymbol{\xi} \text{ inside } V, \\ \frac{1}{2} & \boldsymbol{\xi} \text{ on smooth } S, \\ \frac{1}{4\pi}\Omega & \boldsymbol{\xi} \text{ at corner of } S \text{ with solid angle } \Omega. \end{cases}$$

Hence setting $\boldsymbol{\xi}$ to a smooth point of the surface $S$, one has the boundary integral equation

$$\tfrac{1}{2}\phi(\boldsymbol{\xi}) = \int_S \left( \phi \frac{\partial G}{\partial n} - \frac{\partial \phi}{\partial n} G \right) dS(\mathbf{x}).$$

In this integral equation, one might be given $\phi$ on the surface and have to solve the equation to find $\partial \phi / \partial n$ on the surface, or one might be given $\partial \phi / \partial n$ on the surface and have to solve for $\phi$ on the surface. Once $\phi$ and $\partial \phi / \partial n$ are known on the surface, one can go back to the preceding equation for a point $\boldsymbol{\xi}$ in the interior to find $\phi$ inside $V$.

For external problems, the surface $S$ will have parts on the internal bodies and a part on a surface 'at infinity'. The contribution of the integral over this surface at infinity will be the imposed functional form at infinity, say $\phi_\infty(\mathbf{x})$. Hence one can add this to the boundary integral equation above in place of the integral over the part of the surface at infinity.

### 12.1.1 Greens functions

It is normal to use the 'free-space' Greens functions which do not satisfy any boundary conditions on the surface $S$, i.e. one uses

$$\text{in } R^3: \quad G = -\frac{1}{4\pi|\mathbf{x} - \boldsymbol{\xi}|}, \quad \frac{\partial G}{\partial n} = \frac{(\mathbf{x} - \boldsymbol{\xi}) \cdot \mathbf{n}(\mathbf{x})}{4\pi|\mathbf{x} - \boldsymbol{\xi}|^3},$$

$$\text{and in } R^2: \quad G = \frac{1}{2\pi} \ln |\mathbf{x} - \boldsymbol{\xi}|, \quad \frac{\partial G}{\partial n} = \frac{(\mathbf{x} - \boldsymbol{\xi}) \cdot \mathbf{n}(\mathbf{x})}{2\pi|\mathbf{x} - \boldsymbol{\xi}|^2}.$$

Integrating around the axis for axisymmetric problems, the Greens function can be expressed in terms of elliptic functions. In very simple geometries, it can be worth using images in order to make the Greens function satisfy a boundary condition.

### 12.1.2 Eigensolutions

The boundary integral equation for an interior problem has one eigensolution

$$\phi = 1 \quad \text{and} \quad \frac{\partial \phi}{\partial n} = 0 \quad \text{on } S,$$

corresponding to the trivial constant solution

$$\phi(\mathbf{x}) \equiv 1 \quad \text{in } V.$$

Associated with this eigensolution is the constraint of no net flux

$$\int_S \frac{\partial \phi}{\partial n} \, dS = 0,$$

which follows from zero volume sources in $\nabla^2 \phi = 0$ in $V$.

### 12.1.3 Singular integrals

For $\boldsymbol{\xi}$ fixed at one position on the surface $S$ and $\mathbf{x}$ moving on $S$ in the surface integral,

$$G \propto \frac{1}{|\mathbf{x} - \boldsymbol{\xi}|} \quad \text{in } R^3, \qquad G \propto \ln|\mathbf{x} - \boldsymbol{\xi}| \quad \text{in } R^2.$$

These integrands are singular, although integrable. Hence care must be taken in numerically evaluating these integrals of the Greens function $G$.

The integrals of the normal gradient of the Greens function $\partial G/\partial n$ look more singular. However on a smooth surface, the normal $\mathbf{n}$ is perpendicular to the tangential direction in which $\mathbf{x}$ approaches $\boldsymbol{\xi}$, so

$$\mathbf{n}(\mathbf{x}) \cdot (\mathbf{x} - \boldsymbol{\xi}) \sim \tfrac{1}{2}\kappa|\mathbf{x} - \boldsymbol{\xi}|^2,$$

where $\kappa$ is the curvature. Hence

$$\frac{\partial G}{\partial n} \sim \frac{\kappa}{8\pi|\mathbf{x} - \boldsymbol{\xi}|} \quad \text{in } R^3, \qquad G \propto \frac{\kappa}{4\pi} \quad \text{in } R^2.$$

Hence these integrands are no more singular, although one needs a smooth surface.

## 12.2 Discretisation

There are three steps in turning the boundary integral equation into a discrete numerical version.

- First, the surface must be divided up into 'panels'. In two dimensions this would be the curve divided into segments. In three dimensions, the surface would normally be divided into triangles, but sometimes into rectangles.
- Second, the unknowns $\phi$ and $\partial\phi/\partial n$ must be represented by basis functions $f_i(\mathbf{x})$ over the panels. One can use piecewise constants or piecewise linear functions. For curves, $B$-splines can be used for higher accuracy. Thus on $S$

$$\phi(\mathbf{x}) = \sum \Phi_i f_i(\mathbf{x}), \qquad \frac{\partial\phi}{\partial n} = \sum D\Phi_i f_i(\mathbf{x}),$$

with unknown amplitudes $\Phi_i$ and $D\Phi_i$.
- Third, the integral equation is satisfied, normally at collocation points, but sometimes by a least-squares fit or with weighted integrals. Suitable collocations points are the centre of the panels for piecewise constant basis functions and at the vertices of the panels for piecewise linear basis functions.

One thus forms a discretised version of the integral equation in terms of the amplitudes $\Phi_i$ and $D\Phi_i$

$$\left(\tfrac{1}{2}I - D\mathcal{G}\right)\Phi = -\mathcal{G}D\Phi,$$

where the matrix elements are

$$D\mathcal{G}_{ij} = \int_S f_j(\mathbf{x})\frac{\partial G}{\partial n}(\mathbf{x},\boldsymbol{\xi})\,dS(\mathbf{x}), \quad \text{and} \quad \mathcal{G}_{ij} = \int_S f_j(\mathbf{x})G(\mathbf{x},\boldsymbol{\xi})\,dS(\mathbf{x}),$$

both evaluated at $\boldsymbol{\xi} = \mathbf{x}_i$.

### 12.2.1 Evaluation of the matrix elements

Note that each of the matrix elements $D\mathcal{G}$ and $\mathcal{G}$ involve integration only over a small part of the surface $S$ because each basis function is nonzero only over a few panels. This desirable property is possible when using splines as basis functions only if the compact support $B$-splines are used.

The integrands are singular at the collocation points $\mathbf{x} = \boldsymbol{\xi}$, although analytically integrable. Integrating over panels some distance away from the collocation point, one can use the trapezoidal integration rule with adequate accuracy, at least for the piecewise linear basis functions. A little more accurate would be a Gaussian quadrature.

The singular integrand becomes a problem in the self panel where the integration point is in the panel containing the collocation point. It has been argued that Gaussian quadrature sidesteps the singularity because the Gaussian sample points will avoid the collocation point. However such integration is very inaccurate. Even in the next-to-self panel the nearby singularity leads to poor accuracy. For example eight-point Gaussian quadrature will give an error of $3\ 10^{-15}$ for the nonsingular $\int_0^\pi \sin x\,dx$, while the error is $9\ 10^{-3}$ for the integrable but singular $\int_0^1 \ln x\,dx$.

The correct approach is to *subtract off* the singularity and evaluate that part analytically. Thus in two dimensions, the asymptotic form of approach to the singularity, $x \to \xi$, is

$$G(x, \xi) \sim a(\xi) \ln |x - \xi| + \text{regular term.}$$

Integrating the singularity across the self panel,

$$\int_{\xi-\delta_1}^{\xi+\delta_2} a(\xi) \ln |x - \xi|\,dx = a(\xi)\,(\delta_2 \ln \delta_2 - \delta_2 + \delta_1 \ln \delta_1 - \delta_1).$$

The remaining part is regular and so can be safely integrated numerically, say by the trapezoidal rule. It is wise to use this same subtraction on the next-to-self panel, if not one more beyond.

### 12.2.2 Avoiding the eigensolution

The discretised boundary integral equation for the amplitudes will have singular matrices $DG$ and $G$ because of the eigensolution. One can invert the singular matrices working in the space orthogonal to the eigensolution, although this is a little fiddly. Alternatively there are two fixes.

The first fix is to rely on the truncation error to keep the matrices nonsingular. This is pretty dangerous, but can work.

The second, and much preferred, fix is to make the eigenvalue $\alpha$ rather than 0 for the eigenvector $e$. Thus one replaces singular matrix $A$ by nonsingular matrix $A'$,

$$A' = A + \alpha e e^\dagger,$$

where $e^\dagger$ is the adjoint eigenvector. For the interior problem discussed earlier

$$e = (1, 1, \dots, 1) \quad \text{and} \quad \left(e^\dagger\right)_j = \int_S f_j\,dS.$$

This supposes that the basis functions sum to the constant function, $\sum f_i(x) \equiv 1$.

### 12.2.3 Tests

Coding up the boundary integral method is sufficiently complex that it is well worth testing the program before using it. For the Laplace equation, one has exact solutions for any geometry. Thus in two dimensions

$$\phi = r^k \cos k\theta, \quad \text{with} \quad \frac{\partial \phi}{\partial n} = \mathbf{n} \cdot \nabla\phi = n_r k r^{k-1} \cos k\theta - n_\theta k r^{k-1} \sin k\theta,$$

and similarly in three dimensions.

These exact solutions should nearly satisfy the discretised version of the boundary integral equation. One should test that the error has the correct behaviour, with a second-order error $O(\Delta x^2)$ if piecewise linear basis functions are used for $\phi$ and $\partial\phi/\partial n$. Using cubic splines for the basis functions gives a fourth-order error.

### 12.2.4 Costs

The boundary integral method calculates unknowns on the surface. There are many fewer numerical points on the surface than within the volume. So is the numerical cost much less for the boundary integral method? This question is addressed to solving a Poisson problem. The answer is not obvious.

Consider the Poisson problem inside a two-dimensional square box with $N$ points on a side. There are then $4N$ points on the surface and $N^2$ points in the volume. For the simple square geometry, there are fast Poisson solvers for the problem in the volume which take around $N$ iterations, each iteration of $N^2$ points, and so a cost of $N^3$. The boundary integral equation generates a full $4N \times 4N$ matrix, whereas the volume problem has a very sparse matrix. The cost of inverting a full matrix is $\frac{1}{3}(.)^3$, so a total cost of $21N^3$. This is much the same as the cost of solving the volume problem. In three dimensions, the volume problem would have a cost of $N^4$, while the boundary integral method has a cost of $72N^6$, which is much more expensive.

However, there can be efficiencies if the Poisson problem is resolved every time-step. It is often possible to set up an iteration involving just one matrix evaluation per time-step, at a cost of $(.)^2$. Hence the cost of the boundary integral method can be reduced to $16N^2$ in two dimensions and $36N^4$ in three dimensions, both of which are competitive with solving the volume problem.

These comparisons were for a square box for which there is a fast Poisson solver for the volume problem. The boundary integral method comes into its own in more complex geometries. For exterior problems, the volume method would have to have a grid extending some way towards infinity, while the boundary integral method only has points on the surfaces of the internal bodies. Adaptive grids are easy to deploy for the boundary integral equation by concentrating points on the surface where there is interesting important activity.

## 12.3 Free-surface potential flows

The boundary integral method can be used to solve for the velocity potential $\mathbf{u} = \nabla\phi$ in free surface flows. At the start of a time-step, one knows the position of the surface $S$ and the value of the potential $\phi(\mathbf{x}, t)$ at each point of the surface. The boundary integral method is then used to find the normal gradient of the potential $\partial\phi/\partial n$ at each point on the surface. The tangential gradient of the potential can be found by spatial differentiation of the current values of the potential on the surface. Combining the tangential and normal gradients, one has $\nabla\phi$ at each point on the surface.

The time-step is then made using the instantaneous $\nabla\phi$ on the surface. First points on the surface are moved with the flow,

$$\frac{D\mathbf{x}}{Dt} = \nabla\phi \quad \text{for points } \mathbf{x} \text{ on } S.$$

Second the surface potential is updated,

$$\frac{D\phi}{Dt} = \tfrac{1}{2}|\nabla\phi|^2 - \mathbf{g} \cdot \mathbf{x} - \frac{\gamma}{\rho}\kappa - p_{\text{atm}} \quad \text{for points } \mathbf{x} \text{ on } S,$$

where $\mathbf{g}$ is the gravitational acceleration, $\gamma$ surface tension, $\rho$ density, $\kappa$ the curvature of the surface and $p_{\text{atm}}$ atmospheric pressure.

Because of the existence of capillary waves, there is a restriction on the size of the time-step

$$\Delta t < \sqrt{\frac{\rho}{\gamma}}\Delta x^{3/2}.$$

A good test of a code is the vibration frequencies of an isolated drop.

Free-surface potential flows conserve energy. This leads to an accumulation of numerical noise in the form of short capillary waves. This noise can be removed either by smoothing the surface from time to time, or better by applying a Fourier filter in which the surface displacements are Fourier transformed and then the high-frequency components are cutting out.

## 12.4 Stokes flows

The boundary integral method has been described so far in terms of solving the Laplace equation. The method can also be applied to Stokes flows, which are not much more than a complicated vector version of the Laplace equation. The boundary integral equation for a point on a smooth surface $S$ is

$$\tfrac{1}{2}\mathbf{u}(\xi) = \int_S \left((\sigma \cdot \mathbf{n}) \cdot \mathbf{G} - \mathbf{u} \cdot \mathbf{K} \cdot \mathbf{n}\right) dS(\mathbf{x}),$$

with the Greens function, called a Stokeslet, and its derivative

$$\mathbf{G} = \frac{1}{8\pi\mu}\left(\mathbf{I}\frac{1}{r} + \frac{\mathbf{r}\mathbf{r}}{r^3}\right) \quad \text{and} \quad \mathbf{K} = -\frac{3}{4\pi}\frac{\mathbf{r}\mathbf{r}\mathbf{r}}{r^5}, \quad \text{where} \quad \mathbf{r} = \mathbf{x} - \boldsymbol{\xi}.$$

In the boundary integral equation, the velocity $\mathbf{u}$, stress $\sigma$ and unit normal $\mathbf{n}$ are all functions of $\mathbf{x}$.

To tackle simultaneously the Stokes flow inside and outside a drop, one multiplies the separate integral equations for the inside and outside by the viscosities of the fluids inside and out, $\mu_{in}$ and $\mu_{out}$. Then adding the two equations, one has an integral equation for the velocity on the interface driven by the jump from the outside to the inside in the surface traction $[\sigma \cdot \mathbf{n}] = -\gamma\kappa\mathbf{n}$, with surface tension $\gamma$ and curvature $\kappa$,

$$\tfrac{1}{2}(\mu_{in} + \mu_{out})\mathbf{u}(\boldsymbol{\xi}) = \int_S (\,[\sigma \cdot \mathbf{n}] \cdot \mathbf{G} - (\mu_{in} - \mu_{out})\mathbf{u} \cdot \mathbf{K} \cdot \mathbf{n})\,dS(\mathbf{x}),$$

where $\mathbf{n}$ is the unit normal out of the drop, and now $\mathbf{G}$ does not have a viscosity in the denominator.

For the interior problem, there are eigensolutions of rigid body motion which produce no stress, and a constant normal pressure which produces no flow. These eigensolutions can be used to regularise the singular integrals, e.g. subtracting the curvature $\kappa$ at the collocation point $\boldsymbol{\xi}$ on $S$,

$$\int_S (\kappa(\mathbf{x}) - \kappa(\boldsymbol{\xi}))\mathbf{n} \cdot \mathbf{G}\,dS(\mathbf{x}),$$

leaves the integral unchanged but makes the integrand nonsingular.

For flows in periodic boxes, both Stokes flows and potential flows, the Greens function can be evaluated using Ewald sums. In time evolving flows, a major economy is made by first finding the Greens function at different shears before calculating any flow. The Greens function is singular, so one should split it into a simple singular term plus a smooth remainder, the smooth remainder needing only a small table for interpolation.

# Further reading

*Numerical multipole and boundary integral equation techniques in Stokes Flow* by S. Weinbaum, P. Ganatos and Z.-Y. Yan in Annu. Rev. Fluid Mech. (1990) **22**, 275–316.

# 13

# Fast Poisson solvers

## 13.1 Multigrid method

The multigrid method is a fast Poisson solver for simple geometries, probably the fastest. It will be described for finite differences in a two-dimensional square domain with the number of points $N$ on a side being a power of 2, $N = 2^m$. It is relatively straightforward to extend the method to three dimensions, to the numbers of points which are powers of other small numbers, to cylindrical geometry, with some ingenuity to finite elements, and to other governing equations with problematic long-range behaviour.

As first discussed in §2.3, the Poisson problem for a square of $N \times N$ points has $N^2$ equations in $N^2$ unknowns. Direct inversion of the $N^2 \times N^2$ matrix takes $\frac{1}{3}N^6$ operations, and hence one is interested in faster iterative solution. The simplest iterative method is Gauss–Seidel. To converge one needs to make $O(N^2)$ sweeps through the grid of $N^2$ points, i.e. $O(N^4)$ operations. The small modification to successive over relaxation reduces the number of sweeps required to $O(N)$, i.e. $O(N^3)$ operations. The multigrid method reduces the number of operations to just $O(N^2)$, and with a small numerical coefficient.

The reason that Gauss–Seidel iteration takes $O(N^2)$ sweeps is that the iteration is a little like solving a diffusion equation, where the time for information to diffuse across the grid is proportional to the square of the linear size of the grid. Short length scale errors diffuse away rapidly with Gauss–Seidel iteration. The problem is the slow diffusion of errors varying on the longest length scale. The idea behind the multigrid method is to tackle these long length-scale errors on a coarse grid which has faster diffusion. Better than a two grids, one fine and one coarse, is to have a hierarchy of nested grids. The coarsest grid will have just one interior point and a grid size $\Delta x = \frac{1}{2}$. Subsequent grids have twice the resolution, $(2^k - 1)^2$ interior points and a grid size $\Delta x = \frac{1}{2^k}$. The finest grid has $(2^m - 1)^2$ interior points and a grid size of $\Delta x = \frac{1}{2^m}$.

In order to describe the multigrid method of using the hierarchy of grids, the Poisson problem will be written as a series of linear algebra problems, to find the solutions $x_k$ of

$$A_k x_k = b_k,$$

for grids $k = m$, the finest, to $k = 1$, the coarsest. One makes several V-cycles. Each cycle starts at the finest, descends one level at a time to the coarsest and then ascends back to the finest. For the first cycle, one might as well start the iteration with $x_m = 0$. For subsequent cycles, one starts with $x_m$ the result of the previous V-cycle.

### 13.1.1 A V-cycle

The descending first part of a V-cycle starts with the finest grid, $k = m$.

- One first makes a couple of Gauss–Seidel iterations of $A_k x_k = b_k$. This produces an approximate solution $x_k^{\text{approx}}$.
- This approximate solution is stored for use in the later ascending part of the V-cycle.
- The residue is calculated

$$\text{res}_k = b_k - A_k x_k^{\text{approx}}.$$

- The residue is then coarsened to form the forcing on the next coarser grid

$$b_{k-1} = C_k \text{res}_k \quad \text{where} \quad C_k = \tfrac{1}{16} \begin{pmatrix} 1 & 2 & 1 \\ 2 & 4 & 2 \\ 1 & 2 & 1 \end{pmatrix}.$$

Here the central contribution $\frac{1}{16}4$ is at a point on both the coarse and the fine grids, whereas the neighbouring contributions come from points on the fine grid in between the points on the coarse grid.

- The forcing $b_{k-1}$ is stored for use in the later ascending part of the V-cycle.
- For starting the Gauss–Seidel iterations, $x_{k-1}$ is zeroed.
- The grid level $k$ is reduced by one to a coarser grid.
- If one is now not on the coarsest grid, i.e. $k > 1$, one returns to the top of this list to make a couple of Gauss–Seidel iterations.

The descending part of the V-cycle ends on the coarsest grid with just one internal point. On this grid, the problem $A_1 x_1 = b_1$ is one equation in one unknown and so is solved exactly by a single Gauss–Seidel iteration.

The second ascending part of the V-cycle is now made, starting with $k = 2$.

- The solution $x_{k-1}$ is interpolated to form a correction on the next finer grid

$$x_k^{\text{correction}} = I_k x_{k-1} \quad \text{where} \quad I_k = \frac{1}{4} \begin{pmatrix} 1 & 2 & 1 \\ 2 & 4 & 2 \\ 1 & 2 & 1 \end{pmatrix}.$$

Here the contribution $\frac{1}{4}4 = 1$ is at a point on both the coarse and fine grids. A point on the finer grid in between two points on the coarser grid receives two contributions of $\frac{1}{4}2$ from those two coarser points, whereas a finer point surrounded by four neighbouring coarser points receives contributions of $\frac{1}{4}1$ from those coarser points.

- This correction $x_k^{\text{correction}}$ is added to the approximate solution of this level $x_k^{\text{approx}}$ stored on the descending part of the V-cycle

$$x_k^{\text{better approx}} = x_k^{\text{correction}} + x_k^{\text{approx}}.$$

- Using the stored $b_k$, a couple of Gauss–Seidel iterations of $A_k x_k = b_k$ are now made, starting from $x_k^{\text{better approx}}$, to produce an even better approximation. These Gauss–Seidel iterations diffuse away short length-scale errors produced by the linear interpolation.

- If one is at the finest grid $k = m$, one has completed the V-cycle. Otherwise the grid level is increased by one to a finer grid, and one returns to the top of this list to interpolate the solution onto a finer grid.

It should be noted that the multigrid method is not as simple as starting with the original Poisson problem on the coarsest grid, and then interpolating its solution onto the next finer grid as a starting iteration for the original Poisson problem on that next grid, etc. An important part of the multigrid method is the coarsening of the forcing from a finer grid. This coarsening results in a forcing different from that of the original Poisson problem at that level. This difference corrects the error in the finite difference approximation to the second derivative, an error of about 30% on the coarsest grid.

### 13.1.2 Accuracy and costs

To illustrate the multigrid method, the standard Poisson problem

$$\nabla^2 \psi = -2\pi^2 \sin(\pi x) \sin(\pi y)$$

has been solved in a unit square box, with boundary conditions of $\psi$ vanishing on the boundary. Figure 13.1 plots the maximum value of the residue over the finest grid at the end of a V-cycle as a function of the number of V-cycles taken. The error in the solution is likely to be about $1/2\pi^2$ times smaller. The results are for a finest grid of 256 points on a side, although the residual on the finest grid was found to be the same when only 64 points were used.

The residual is reduced by roughly the same factor by each V-cycle. With two Gauss–Seidel iterations at each level during descending and during ascending, the residual is reduced by a factor of 10. With three iterations at each level, the residual is reduced by 15, while with just a single iteration it is reduced by about 4.

The coarsening operation and the interpolation both involve visiting every point in the current grid, just as in one Gauss–Seidel iteration. Hence to compare the numerical cost, the number of V-cycles taken has been multiplied by the number of Gauss–Seidel iterations at each level plus one. Figure 13.1 shows that this numerical cost is much the same whether one, two or three iterations are made, with a small advantage for taking two iterations.

The number of operations on the finest grid of $N^2$ points for the descending or ascending parts will be $3N^2$ for two Gauss–Seidel iterations plus either a coarsening or an interpolation. The number of operations on the next coarser grid will be $\frac{1}{4}$ of this, and $\frac{1}{16}$ on the next grid. Summing the geometric series and the descending and ascending parts, one estimates the number of operations per V-cycle as $8N^2$.

To obtain a solution with an accuracy of $10^{-4}$, the residue has to be less than $2\,10^{-3}$, which requires just four V-cycles with two Gauss–Seidel iterations, i.e. $32N^2$ operations. This should be compared with $2N^3$ operations using SOR. Thus on a $128 \times 128$ grid, the multigrid method is 32 times faster, well worth the extra coding effort.

## 13.2 Fast Fourier Transforms

Chapter 6 discussed spectral methods, which are useful for very smooth functions in very simple geometries. As noted in §6.2, the Poisson problem becomes trivial in Fourier space. However, there is a significant cost in transforming a Poisson problem in real space to a trivial problem in Fourier space and then transforming back to a solution in real space.

For an $N \times N$ problem in two dimensions, the $N^2$ Fourier amplitudes, each a sum of $N^2$ terms, looks to be an $O(N^4)$ method of solving the Poisson problem, thus comparable with Gauss–Seidel iteration. Applying the Orszag speedup of reusing partial sums as described in §6.9, the number of operations is reduced from $O(N^4)$ to $O(N^3)$, thus comparable with SOR iteration. Finally using Fast Fourier Transforms (FFTs) of §6.9 reduces the operation count to $O(N^2 \ln N)$,

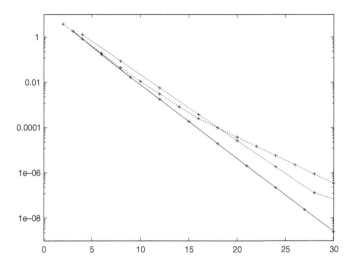

Figure 13.1 The residual as a function of the numerical cost calculated as the number of V-cycles taken multiplied by the number of Gauss–Seidel iterations at each level plus one. The lowest curve has two Gauss–Seidel iterations at each level, the middle curve at the right-hand end has three iterations and the top just one.

which is just a little more expensive than the $O(N^2)$ of the multigrid method. Note, however, the fast transform is really only available for Fourier transforms and so is restricted to Cartesian geometry with fully periodic boundary conditions.

Some problems are fully periodic in some, but not all, directions. Hence one might use a combination of spectral methods for the periodic directions and finite differences in real space for the nonperiodic directions. For example, to simulate turbulent flow in a channel, one could use Fourier modes in the downstream and spanwise directions and finite differences for the cross-channel direction. The Poisson problem would then reduce to inverting a tridiagonal matrix in the cross-channel direction for the finite differences of Fourier amplitudes of the two periodic directions. Inverting a tridiagonal matrix is a fast $O(N)$ operation, so solving the three-dimensional Poisson problem would cost $O(N^3 \ln N)$ if using FFTs in the two periodic directions. Note however, differentiation using finite differences is considerably less accurate than the special spectral accuracy of spectral methods. One might therefore wish to use the higher-order finite differences of §4.1 or the higher-order compact solver of §4.2.

## 13.3 Domain decomposition

While the multigrid method and Fast Fourier Transforms are restricted to very simple geometries, essential Cartesian, domain decomposition can be applied to complex geometry, e.g. to flow past an aircraft with complex wings and engine mountings. Domain decomposition is also good for very large problems, again, for example, the flow past an aircraft, which cannot be easily fitted into the memory of a computer. Domain decomposition is readily parallelisable. Domain decomposition is applicable to finite differences and finite elements. A version of finite elements is Finite Element Tear and Interconnect (FETI). With finite elements, the unstructured grid is given sufficient structure by the geometry of which points are near others.

The idea is to divide the domain into a number, possibly a large number, of subdomains. For each subdomain, there will be internal points and boundary points, and associated with the points, internal and boundary variables. In finite differences, the variables are the values at the points. In finite elements, the variables are values of the amplitudes of the basis functions. Associated with the variables, there will be governing equations. It is important that the governing equation for an internal point only involves internal and boundary variables of that subdomain. On the other hand, the governing equation for a boundary point will involve variables from the subdomains on both sides of the boundary, and more sides at a corner.

The domain decomposition method proceeds in each subdomain by solving for the internal variables in terms of the boundary variables. The problem in each subdomain is thus independent, so can be simultaneously tackled on separate processors, each with only a fraction of the total memory. With all the internal variables now known, the problem for the boundary variables on the internal interfaces between subdomains can be tackled. If this problem is too large, the subdomains can be grouped into regions, and the variables on the interfaces within the regions first tackled, again parallelisable, before finally solving for the variables on the interfaces between the regions. The equations for the boundary variables on the interfaces between subdomains will involve the values of nearby internal variables, whose values are by now known in terms of various boundary variables.

To give a mathematical expression to the above recipe, write the Poisson problem as the linear algebra problem to find the solution $x$ of $Ax = b$. Let the domain be divided into $K$ subdomains. Partition the solution $x$ into the internal variables of the different subdomains, $x_1, x_2, \ldots, x_K$, and the boundary variables $y$. Now the internal variables of one subdomain are isolated from the internal variables of all the other subdomains, and so are governed by

$$A_k x_k + B_k y = b_k.$$

On the other hand, the problem for the boundary variables involves all the internal variables, i.e.

$$C_1 x_1 + C_2 x_2 + \cdots + C_K x_K + Dy = b_0.$$

In other words, the original problem $Ax = b$ has been turned into the nearly block-diagonal form

$$\begin{pmatrix} A_1 & 0 & & 0 & B_1 \\ 0 & A_2 & & 0 & B_2 \\ \vdots & \vdots & \ddots & & \vdots \\ 0 & 0 & & A_K & B_K \\ C_1 & C_2 & \cdots & C_K & D \end{pmatrix} \begin{pmatrix} x_1 \\ x_2 \\ \vdots \\ x_K \\ y \end{pmatrix} = \begin{pmatrix} b_1 \\ b_2 \\ \vdots \\ b_K \\ b_0 \end{pmatrix}.$$

The internal variables are found as

$$x_k = A_k^{-1}(b_k - B_k y).$$

These $K$ problems, one for each subdomain, are isolated from one another, so they can be solved in parallel. Substituting the solutions for the internal variables into the problem for the boundary variables produces what is known as the 'Schur complement system',

$$\left(D - C_1 A_1^{-1} B_1 - \cdots - C_K A_K^{-1} B_K\right) y = b_0 - C_1 A_1^{-1} b_1 - \cdots - C_K A_K^{-1} b_K.$$

Solving for the unknown variables can mean one of several alternatives. If the subdomains are sufficiently small, then direct solution, say by LU decomposition, would not be too expensive. For larger subdomains, and for the Schur complement for the boundary variables, one might consider an iterative solution, say by conjugate gradients.

Domain decomposition can be interpreted in several ways. Finding the inverse matrix for each subdomain can be thought of as finding a particularly good preconditioner (see §8.4) for the matrix for the full domain, i.e. the diagonal preconditioner

$$P = \begin{pmatrix} A_1^{-1} & & & & \\ & A_2^{-1} & & & \\ \vdots & \vdots & \ddots & & \vdots \\ & & & A_K^{-1} & \\ & & \cdots & & I \end{pmatrix},$$

reduces the problem $Ax = b$ to

$$\begin{pmatrix} I & & & & A_1^{-1}B_1 \\ & I & & & A_2^{-2}B_2 \\ \vdots & \vdots & \ddots & & \vdots \\ & & & I & A_K^{-1}B_K \\ C_1 & C_2 & \cdots & C_K & D \end{pmatrix} \begin{pmatrix} x_1 \\ x_2 \\ \vdots \\ x_K \\ y \end{pmatrix} = \begin{pmatrix} A_1^{-1}b_1 \\ A_2^{-1}b_2 \\ \vdots \\ A_K^{-1}b_K \\ b_0 \end{pmatrix}.$$

The boundary variables can be assigned as independent variables to both the adjacent subdomains, and then requiring them to take the same value becomes a Lagrange multiplier in the global problem. In terms of the Poisson problem, the equation for the boundary variables is equivalent to choosing the boundary value so that the normal flux is continuous across the boundary. Expressing internal variables in terms of boundary values is similar to the boundary element method without a priori knowing the Greens function.

There are versions of the method which have overlapping subdomains. Indeed the first ideas date back to 1870 when Schwarz tackled the Poisson problem in a keyhole-shaped domain which was the intersection of a circle and a rectangle. The problem was tackled iteratively, solving analytically for one domain using values on its boundary from the interior on the previous solution of other domain.

### 13.3.1 Costs

If using direct LU inversion, domain decomposition yields significant savings. Solving the full domain $N \times N$ problem in two dimensions requires $N^6$ operations. Dividing the domain into $K$ subdomains reduces the operations to $N^6/K^3$ in each subdomain, and the subdomains can be processed in parallel. There would remain about $NK^{1/2}$ boundary points in the Schur complement, requiring $N^3K^{3/2}$ operations. For example for $N = 10^2$ and $K = 25$, the full domain requires $10^{12}$ operations, while each subdomain and the Schur complement require only $10^9$ operations. Solving the full domain $N \times N \times N$ in three dimensions requires $N^9$ operations, which reduces to $N^9/K^3$ in each subdomain and $N^6K$ for the Schur complement. For example for $N = 10^2$ and $K = 27$, the full domain requires $10^{18}$ operations, while each subdomain and the Schur complement require only $10^{14}$ operations. This is a significant saving in the number of operations, as well as a reduction in the size of memory required.

# 14

# Fast Multipole Method

The Fast Multipole Method is in effect a fast Poisson solver for a particular type of problem, one with discrete particles rather than a continuously distributed source.

Potential flow and Stokes flow produce long-range interactions between particles in a fluid. Calculating the sum of all the long-range interactions between all $N$ particles would seem a big $O(N^2)$ problem. Fortunately there are clever tricks which reduce size of the problem considerably. By clustering the effects of distant particles into effective monopoles, the Barnes–Hut algorithm reduces the $O(N^2)$ to $O(N \ln N)$. Giving the effective particles more complex multipole structure and representing their induced field within a neighbourhood by low-order polynomials, the Greengard–Rokhlin algorithm reduces the size to just $O(N)$. The latter $O(N)$, however, comes with a large coefficient, $O(36 \ln^2 1/\epsilon)$ in two dimensions and $O(189 \ln^4 1/\epsilon)$ in three dimensions for accuracy $\epsilon$, so that is only worth the complex coding when $N$ is very large, typically larger than $10^4$ and $10^6$ respectively.

The Fast Multipole Method has applications beyond fluid mechanics, to gravitational interactions in astrophysics, electromagnetic wave interactions and density functional theory. The ideas have also been used in calculations with large matrices which have elements which decay away from the diagonal. The methods will be presented here for the case of a two-dimensional complex potential

$$w(z_i) = \sum_{j \neq i}^{N} q_j \ln(z_i - z_j),$$

for the potential $w$ at $z = z_i$ due to charges $q_j$ at $z = z_j$ inside a square box. For the motion of point vortices, one might be interested in the gradient $\nabla w$ instead of the value $w$. The generalisations to three dimensions and to Stokes flow are relatively straightforward.

## 14.1 Trees, roots and leaves

The method uses a hierarchy of subdomains. The initial square box is first divided into four equal squares. Each of these squares is then divided into four equal smaller squares. This process of repeated subdivision continues through $\log_4 N$ levels, so that on average there is one particle in each of the boxes of the smallest size. Of course some of these smallest boxes will have more than one particle, and some will have none. If the initial domain was not a square but a long rectangle, then the first generation of subdivision might divide the longer size into more than two in order to make a squarish first subdivision. If the particles are very unevenly distributed within the box, it may be useful to subdivide beyond $\log_4 N$ levels, so that there are few particles in all the smallest boxes.

For a box at any one level, a smaller box of the next level within it is called a child, while the larger box of the next level which contains it is called the parent. There is thus a tree structure and coding the method is very much about moving up and down the tree. The top of the tree is the original domain, called the root. Once a branch of the tree contains no particles, there is no need to continue to subdivide that box. The smallest box down a branch that contains a particle is called a leaf.

## 14.2 Barnes–Hut algorithm

This algorithm uses just monopoles. One starts with an upward pass, starting at the leaves and proceeding to the root. This is followed by $N$ downward passes, one for each particle.

For the *upward pass*, the charges $q_c$ of the children are summed to find the charge of the parent, $q_p = \sum q_c$. In addition, the centre of mass of the charges of the children is calculated, and this becomes the position of the parent charge, $z_p = \sum z_c q_c / \sum q_c$. This upward pass starts at the leaves and moves up the tree to the level just below the root, clustering the contributions of the children into a parent one level at a time.

During the downward pass for one particle, there is a question of whether or not a particular box is adjacent to the box of the same size which contains the particle. If it is not adjacent, it is measured as *far*. This is illustrated in Figure 14.1 with a particle at $(56, 24)$ in an $80 \times 80$ box. At the first level of subdivision, all four of the $40 \times 40$ boxes either contain the particle or are adjacent to the one that does, so that there are no *far* boxes. At the second level of subdivision, 7 of the 16 $20 \times 20$ boxes are not adjacent to

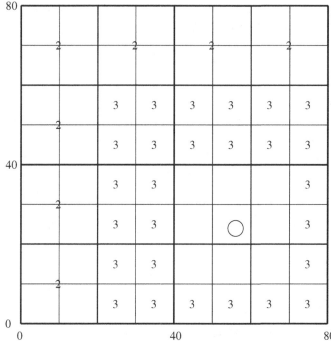

Figure 14.1 The circle represents a particle at $(56, 24)$. The $20 \times 20$ boxes marked with a 2 are *far* from the particle at the second level of subdivision, while the $10 \times 10$ boxes marked with a 3 were in $20 \times 20$ boxes which were *not far* at the second level of subdivision but are *far* at the third level.

the $20 \times 20$ box containing the particle, and so are *far*. These are marked with a 2 in Figure 14.1. At the third level of subdivision, five of the remaining nine $20 \times 20$ boxes each containing four $10 \times 10$ boxes are now all *far* from the particle, plus there are seven new $10 \times 10$ boxes in the lower-right corner which are *far*. These are all marked with a 3 in Figure 14.1. In fact away from boundaries, there are always 27 boxes which newly become far, coming from the 8 parent-boxes adjacent to the parent.

Consider the *downward pass* for the $j$th particle at $z_j$. If at some level a box $b$ is measured as *far* from the particle, then its contribution to the potential is calculated from the clustered charge $q_b$ at $z_b$ of that level,

$$q_b \ln(z_j - z_b),$$

and one does not descend further down that branch of the tree. If at some level a box is measured as *not far* and it is not at the lowest level, then one goes down a level to the four children below. On the other hand, if a box is measured as *not far* and one is at the lowest level, then one sums the contributions from the individual particles within that box,

$$\sum_{i \text{ in box}} q_i \ln(z_j - z_i).$$

Note that the tree structure has the effect of more distant particles being combined into larger clusters.

The Barnes–Hut algorithm is an $O(N \ln N)$ process, as reasoned now. There are $N$ downward passes, one for each particle. Each downward pass moves through $\log_4 N$ levels. At each level, there are 27 new boxes which become *far*, fewer if near the boundary. Hence there at most $27N \log_4 N$ interactions compared with the naive calculation of $N^2$ interactions. Thus the algorithm only becomes less work if the number of particles exceeds 200. That estimate is for a square domain in two dimensions. For a cube domain in three dimensions, there are 26 adjacent boxes rather than 8, each of which contain 8 children rather than 4. The number of interactions thus becomes $189N \log_8 N$, which is less than the direct $N^2$ only if the number of particles exceeds 2,000.

## 14.3  Fast Multipole algorithm

The Barnes–Hut algorithm achieves its efficiency from clustering the distant particles in their effect on each particle. The Fast Multipole algorithm goes further by calculating the effect of a cluster on a cluster of particles, rather than the effect on individual particles. This requires firstly giving more structure to the clusters through multipoles rather than just the monopoles of Barnes–Hut, and secondly calculating the variation of the field across the receiving cluster rather just its value at a particle.

### 14.3.1  Upward pass

The Fast Multipole algorithm has two passes, first an upward pass from the leaves at the bottom of the tree up to the top subdivisions and then a downward pass from the top to the bottom. At the lowest level in the *upward pass*, the point charges $q_i$ at $z_i$ are expressed as multipoles about the centre of their box at $z_c$ using the *first far shift*

$$\ln(z - z_i) = \ln(z - z_c) + \sum_{r=1}^{\infty} \frac{(z_c - z_i)^r}{r(z - z_c)^r}.$$

At higher levels in the upward pass, the contributions from the multipoles of the children at $z_c$ are re-expressed as multipoles about the centre of the parent at $z_p$. The first far shift above is used to move the monopoles of the children to the parent. The higher multipoles need the *second far shift*

$$\frac{1}{(z - z_c)^m} = \sum_{r=0}^{\infty} b_r^m \frac{(z_p - z_c)^r}{(z - z_p)^{m+r}},$$

where $b_r^m$ is a binomial coefficient. The infinite sums are truncated quite early, with multipoles only to a certain order $m_{max}$ retained. Selecting how many multipoles should be retained depends on the accuracy required, see later.

### 14.3.2 Downward pass

The *downward pass* starts two levels down from the root. At a particular level, all the boxes of that level must be considered before moving down a level. A box will receive two different inputs: it inherits from its parent via a *local shift* input from boxes which were *far* from its parent, and it receives via a *local expansion* input from the 27 newly *far* boxes of its own level which are contained within the parent-boxes adjacent to its parent. At the lowest level, the above inputs are found first. The resulting field is then evaluated at each particle. Then direct particle–particle interactions are added from particles within its own box and the eight immediately adjacent boxes.

The *local shift* moves a polynomial variation centred on a parent at $z_p$ to one centred on the child at $z_c$ by

$$(z - z_p)^m = \sum_{r=0}^{m} c_r^m (z - z_c)^r (z_c - z_p)^{m-r},$$

where $c_r^m$ is a binomial coefficient.

The *local expansions* of the *far* monopole and the *far* higher $m$-multipole at $z_b$ about the centre of the child at $z_c$ are

$$\ln(z - z_b) = \ln(z_c - z_b) + \sum_{r=1}^{\infty} \frac{(z - z_c)^r}{r(z_c - z_b)^r}, \quad \text{and} \quad \frac{1}{(z - z_b)^m} = \sum_{r=0}^{\infty} b_r^m \frac{(z - z_c)^r}{(z_c - z_b)^{m+r}}.$$

The infinite sums are truncated at $r = m_{max}$, thus restricting the polynomial variation in the local variation to at most $(z - z_c)^{m_{max}}$.

### 14.3.3 Errors

The truncation errors come from the first multipole not included, i.e. from terms like

$$\left(\frac{z - z_c}{z_c - z_b}\right)^{m_{max}+1}.$$

Now the farthest the point $z$ in a box can be from the centre of that box $z_c$ is $\sqrt{2}d/2$, where $d$ is the length of the side of the square box. And the shortest separation between the centres of two *far* boxes, $z_c$ and $z_b$, is $2d$. At these two extremes, the first omitted term would be

$$\text{Error} \leq \left(\frac{1}{2\sqrt{2}}\right)^{m_{\max}+1}.$$

For this to be less than $10^{-3}$, one needs to include multipoles and polynomial variations to the order $m_{\max} = 6$. In three dimensions, the farthest a point can be from the centre of its box is $\sqrt{3}d/2$, so that the error estimate becomes $(\sqrt{3}/4)^{m_{\max}+1}$, now requiring $m_{\max} = 8$ for $10^{-3}$ accuracy. Depending on the distribution of particles, the above estimates of the error may be untypically large.

### 14.3.4 Costs

First we count the number of calculations for one particle at the lowest level of the downward pass. With an average of one particle in each box, there will be at the lowest-level eight direct particle–particle contributions to each particle. The local shift has $m_{\max} + 1$ terms in the polynomial to evaluate. Finally there are 27 newly *far* boxes which contribute to the $m_{\max} + 1$ terms in the polynomial from the $m_{\max} + 1$ multipoles. Thus at the lowest level, the number of calculations is

$$8N + (m_{\max} + 1)N + 27(m_{\max} + 1)^2 N.$$

At the next level up there are $\frac{1}{4}N$ boxes, requiring evaluations of the polynomial and contributions from 27 newly *far* boxes. At the level up one more there are $\frac{1}{16}N$ boxes. Summing the geometric series for the contributions from all the levels, one arrives at the number of calculations as

$$8N + \tfrac{4}{3}(m_{\max} + 1)N + 36(m_{\max} + 1)^2 N.$$

Finally the number of multipoles to achieve a required accuracy is given by

$$m_{\max} + 1 = \frac{\ln(1/\text{Error})}{\ln 2\sqrt{2}}.$$

Hence for an error of $10^{-3}$, it is only when the number of particles $N$ exceeds 10,000 that the Fast Multipole Method involves fewer calculations than $N^2$ direct particle–particle interactions.

In three dimensions, the number of nearby boxes is much larger, and there are $m_{\max}^2$ terms in mutipoles up to order $m_{\max}$. These considerations increase the number of calculations to

$$26N + m_{\text{max}}^2 N + 189 m_{\text{max}}^4 N,$$

and hence the algorithm becomes less work at an accuracy of $10^{-3}$ only with well over $10^6$ particles.

# 15

---

# Nonlinear considerations

## 15.1 Finding steady states

For the nonlinear system

$$\dot{\mathbf{u}} = \mathbf{f}(\mathbf{u}),$$

whether this is a discretised version of the Navier–Stokes equations or is a reduced model system, one can search for steady-states by integrating in time from some different initial values. There are several disadvantages to this approach if one is only interested in the eventual steady states. First, the time evolution may be slow and consume unnecessary computer effort. Second, there may be several possible steady states, some of which might be missed with the wrong initial conditions. And finally, a steady state might be unstable, and so never approached from any initial condition, although an unstable steady state might still have some interest. On the other hand if no steady states exist, integrating from any initial condition will give a clear hint of the lack of steady states.

Searching directly for steady states is essentially a root solving problem, which tends to be a very nonlinear problem. Hence one uses iterative methods to home in on a root. First, however, one needs a reasonable first approximation, which may be difficult. One can solve some initial value problems for a short time to see if the solution is heading towards a steady state, although it would necessarily have to be a stable steady state. Once one has a reasonable first approximation, then Newton–Raphson iteration delivers very good accuracy after just a couple of iterations, because the number of accurate decimal places doubles with each iteration. Newton–Raphson iteration is

$$\mathbf{u}_{n+1} = \mathbf{u}_n + \delta_n, \quad \text{where} \quad \mathbf{f}'(\mathbf{u}_n)\delta_n = -\mathbf{f}(\mathbf{u}_n).$$

If the Jacobian $\mathbf{f}'$ vanishes, then the good rate of convergence is lost, and the error just halves at each iteration (if $\mathbf{f}''$ does not vanish).

The Jacobian $\mathbf{f}'$ is additionally useful because it contains information about the stability of the steady states. The steady state would be linearly stable if all the eigenvalues of $\mathbf{f}'$ had negative real parts, $Re(\lambda) < 0$. Finding eigenvalues was discussed in §8.6 and the following sections in Chapter 8. For small systems, it is not necessary to find the values of all the eigenvalues in order to test whether the real part of each of them is negative. One can form the characteristic polynomial satisfied by the eigenvalues and apply the Routh–Hurwitz process of examining various determinants of the coefficients of the polynomial.

### 15.1.1 Finding the Jacobian

The major problem with Newton–Raphson iteration is that one needs the Jacobian, and for a large system the Jacobian would be very large. Very rarely one can form an expression for it analytically. For a system of partial differential equations, there is an issue of whether the analytic expression should be obtained before or after discretisation. The following subsection has an example of a small system in which the Jacobian is formed analytically before discretisation.

If the Jacobian cannot be found analytically, it has to be found numerically. This means numerical differentiation

$$\frac{\partial f_i}{\partial u_j}(\mathbf{u}_n) \approx \frac{f_i(\mathbf{u}_n + h\mathbf{e}_j) - f_i(\mathbf{u}_n)}{h},$$

for a suitable small $h$ and for one direction $\mathbf{e}_j$. Note that in order to find where the vector $f_i(\mathbf{u})$ vanishes, it is necessary to find the derivative in all possible directions $\mathbf{e}_j$.

As finding the Jacobian every iterative step can be expensive, one can use information from the iteration to update the Jacobian with less effort. The update

$$\mathbf{f}'_{n+1} = \mathbf{f}'_n + \frac{2\mathbf{f}_n \delta_n}{|\delta_n|^2}$$

corrects linear variations in $\mathbf{f}'$ in the direction of $\delta_n$. The update should be a small correction, so it may be of little help and can be ignored. If the correction is not small, say because $\mathbf{u}_n$ is far from the steady state, then the update may result in a divergent iteration.

### 15.1.2 Example of the limit cycle of the Van der Pol oscillator

This example is not quite a steady state but rather a steady nonlinear oscillation. The system is small, just a second-order ordinary differential equation, so that an expression for the Jacobian can obtained analytically, and prior to the discretisation. The final evaluation of the Jacobian does however require a numerical integration of an ordinary differential equation.

The van der Pol oscillator is governed by

$$\ddot{u} + \mu\dot{u}(u^2 - 1) + u = 0,$$

with parameter $\mu$. All solutions tend to a nonlinear periodic limit cycle. If the period is $T(\mu)$, the periodic solution will have

$$u(T) = u(0) \quad \text{and} \quad \dot{u}(T) = \dot{u}(0).$$

As the limit cycle oscillates around $u = 0$, one can define the time origin to be when $u$ vanishes, $u(0) = 0$.

One looks at all the solutions $u_{n+1}(t)$ with period $T_{n+1}$ near to an approximation to the limit cycle, $u_n(t)$ with period $T_n$,

$$u_{n+1}(t) = u_n(t) + \epsilon v(t) \quad \text{and} \quad T_{n+1} = T_n + \delta,$$

with $\epsilon$ and $\delta$ small. Linearising around the approximation $u_n(t)$, the variation $v(t)$ satisfies

$$\ddot{v} + \mu\dot{v}(u_n^2 - 1) + \mu\dot{u}_n 2u_n v + v = 0.$$

Without loss of generality, one can take initial conditions $v(0) = 0$ and $\dot{v}(0) = 1$. One solves numerically the equation for $v(t)$ simultaneously with solving numerically the equation for $u_n(t)$. Linearising the periodicity conditions for $u_{n+1}(t)$ with $T_{n+1}$ gives

$$u_n(T_n) + \delta\dot{u}_n(T_n) + \epsilon v(T_n) = u_n(0) + \epsilon v(0) = 0,$$
$$\dot{u}_n(T_n) + \delta\ddot{u}_n(T_n) + \epsilon\dot{v}(T_n) = \dot{u}_n(0) + \epsilon\dot{v}(0) = \dot{u}_n(0) + \epsilon.$$

This pair of linear equations is solved for $\epsilon$ and $\delta$, forced by the failure of the approximation $u_n(t)$ to be strictly periodic $(u_n(T_n) - u_n(0))$ and $(\dot{u}_n(T_n) - \dot{u}_n(0))$. The pair of linear equations is the Jacobian for the problem. The equation governing $v(t)$ is obtained analytically from the original equation, and before discretisation. Solving the equation for $v(t)$ requires numerical input to the final evaluation of the Jacobian.

## 15.2 Parameter continuation

Many problems contain a parameter, say the Reynolds number or $\mu$ in the Van der Pol oscillator above. One may be interested in how the steady state varies with changing the value of the parameter. Consider a problem with a parameter $\alpha$

$$\dot{\mathbf{u}} = \mathbf{f}(\mathbf{u}, \alpha).$$

Let $\mathbf{u}_0$ be the steady solution for the parameter $\alpha_0$. Then making a small increment in the parameter, to $\alpha_0 + \delta\alpha$, one can find a first estimate of the new solution $\mathbf{u}_0 + \delta\mathbf{u}$ from

$$\delta\mathbf{u} \cdot \frac{\partial \mathbf{f}}{\partial \mathbf{u}} + \delta\alpha\frac{\partial \mathbf{f}}{\partial \alpha} = 0.$$

Newton–Raphson iteration can then be used to refine this first estimate.

There is a problem with this continuation, however, when $\partial\mathbf{f}/\partial\mathbf{u}$ becomes singular at some value $\alpha$. At this point the Jacobian has an eigenvalue which vanishes. The steady state thus loses it stability. The eigenvector gives the direction of a new branch of solutions *bifurcating* from the previous family of steady states.

Another common possibility is for the real part of a complex conjugate pair of eigenvalues of the Jacobian to vanish at some value $\alpha$. There is then a Hopf bifurcation, with a pair of new solutions branching away from the previous family of steady states, see Figure 15.1a.

Sometimes two branches of solutions coalesce at a particular value of $\alpha$ and there is no solution for greater values of $\alpha$, i.e. there is a *turning point* as illustrated in Figure 15.1b. With $\mathbf{u}$ multivalued as a function of $\alpha$, one can think of switching the parameter from $\alpha$ to possibly $|\mathbf{u}|$.

Finally, it is possible for the family of steady solutions to simply terminate at a *limit point*, as illustrated in Figure 15.1c.

## 15.3 Searching for singularities of physical problems

Some physical problems seem to have singularities, or more properly, some simplified mathematical models of physical problems can have mathematical singularities. For example, the unsteady incompressible boundary layer equations for an impulsively started cylinder have a finite-time singularity at $t \approx 3.0$, with time nondimensionalised by the velocity and radius of the cylinder. For example, the inviscid two-dimensional vortex sheet has a finite-time

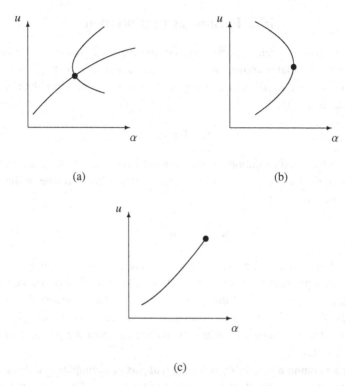

Figure 15.1 Various behaviours when the Jacobian vanishes. (a) A bifurcation in the solution, (b) A turning point in the solution and (c) A limit point in the solution.

singularity in the curvature of the sheet. These singularities come from nonlinearities in the governing equations.

Finite computers are unsuitable for resolving singular behaviour. By taking smaller time-steps and/or by using a finer spatial grid, one can approach nearer and nearer to a singularity, but with a finite resolution one must always fail just before the singularity. Clustering grid points near to a site of interesting behaviour may also help postpone the point of failure, but failure cannot be avoided.

To demonstrate a singular behaviour at a finite time, it is unconvincing to plot naively the quantity that is blowing up as a function of time, as in $A(t)$ in Figure 15.2a. If it is already known from physical arguments that $A(t)$ blows up like say $(t_s - t)^{-1/2}$, then it is better to plot $1/A^2(t)$ as a function of $t$ to see a linear approach to $1/A^2 = 0$, see Figure 15.2b. This plot enables one to find the time of the singularity $t_s$. If one does not know the power-law of the singular

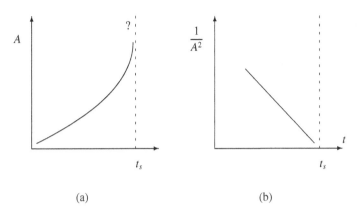

Figure 15.2 Investigating singular behaviour. (a) Possible singular $A(t)$? and (b) Confirmed singular behaviour.

behaviour, one can make a log–log plot of $A(t)$ against $t_s - t$, but for this one needs a good guess for the time of the singularity $t_s$. An alternative way to find the unknown power-law $\alpha$ and unknown time $t_s$ in the singular behaviour $A \sim (t_s - t)^{-\alpha}$, is to plot the logarithmic derivative

$$-\frac{A}{A_t} \sim \frac{1}{\alpha}(t_s - t).$$

However, it is much better to have some theoretical idea of the type of singularity that one is looking for.

At the end of §2.10, the force on the top plate of a driven cavity was investigated at early times. Knowing that the force should behave as

$$F(t) \sim \frac{1}{2}\sqrt{\frac{Re}{\pi t}},$$

Figure 2.16 plotted a scaled force $F\sqrt{t/Re}$ as a function of time $t$, with limited success. From the above discussion, it would be much better to plot $1/F^2$ as a function of time, expecting a linear approach to zero time like $4\pi t/Re$. This is done in Figure 15.3 with a more convincing approach to the theoretical result than in Figure 2.16.

### 15.3.1 Use of computer algebra

In most applications of computers to problems in fluid mechanics, the computer is used to make arithmetic calculations, many arithmetic calculations, reliably and rapidly. In some special problems, the computer can be used differently to make algebraic calculations, and again reliably and rapidly. Using

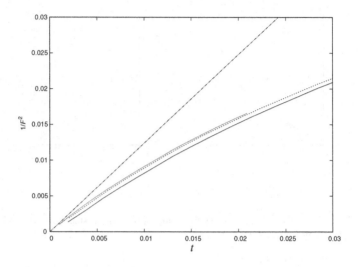

Figure 15.3 The force $F(t)$ on the top plate of the driven cavity at early times. Plotting $1/F^2$ as a function of time $t$ for $N = 80$, 160 and 320. The line corresponds to $4\pi t/Re$.

computer-aided algebra, it is sometimes possible to obtain a large number of terms in a power series expansion of a quantity of interest

$$A(t) \sim \sum_{}^{N} a_n t^n.$$

A Domb–Sykes plot of adjacent coefficients then reveals the time $t_s$ and the power-law $\alpha$ of the singularity $(t_s - t)^{-\alpha}$ which is nearest $t = 0$

$$\frac{a_n}{a_{n-1}} \sim \frac{1}{t_s}\left(1 + \frac{\alpha - 1}{n}\right).$$

Alternatively one can convert the power series by further algebra manipulations into a *Padé approximant*, a quotient of two power series,

$$\sum_{}^{N} a_n t^n = \frac{\sum^K b_n t^n}{\sum^L c_n t^n},$$

with $c_0 = 1$ for normalisation. One can then examine at the zeros of the denominator in order to find the singular behaviour of $A(t)$. One would be interested in zeros, possibly on the complex $t$-plane, which moved little when more and more terms were included, i.e. $K$ and $L$ increased.

# 16

---

# Particle methods

With the exception of Chapter 14 on Fast Multipole Methods, this book has been concerned with numerical methods for a continuum description of fluids by partial differential equations. This chapter switches to particle descriptions. The first four sections treat the fluids, dense liquids or dilute gases, as moving atoms. The following three sections are about macroscopic particles moving in a continuum fluid. In the final section, the continuum fluid is viewed as a collection of fluid blobs, or particles, which interact with adjacent blobs through pressure and friction forces.

## 16.1 Molecular dynamics

In molecular dynamics, pioneered by Alder and Wainwright in 1959, atoms move according to Newton's equation of motion with interatomic forces such as Coulomb and van der Waals. Sometimes more complex forces are used to represent better their quantum mechanical origin. Early calculations were to find the viscosity of inert liquids. Recent calculations examine molecules passing through channels in cell membranes and the folding of proteins. In fluid mechanics, one can study the slip height boundary condition and the dynamics of molecules near the moving contact line of a liquid drop on a solid surface. The number of atoms considered varies from a few hundred to a hundred million. The intrinsic time-step is a few femtoseconds ($10^{-15}$ s). Most phenomena of interest have a time-scale of nano- to microseconds, which means very very many time-steps, which means very long computations, of perhaps several CPU-months. Fortunately, the dynamics is suitable for parallel computing.

In the simplest case, van der Waals forces are calculated from a Lennard–

Jones potential

$$V = 4\epsilon \left( \left( \frac{\sigma}{r} \right)^{12} - \left( \frac{\sigma}{r} \right)^{6} \right),$$

where $\epsilon$ is the interaction energy of a few $kT$, $r$ is the distance between the centres of two atoms and $\sigma$ is their effective atomic diameter. The force is weakly attractive at large separations and becomes very repulsive at separations less than $\sigma$. The force is to be summed over all pairs of atoms. However because of the rapid decay, it can be cut off at a suitable distance. This reduces the number of interactions from $O(N^2)$ to $O(N)$, where $N$ is the number of atoms in the simulation. To achieve this reduction, a list is kept of nearby atoms which might interact. Coulombic electrostatic forces cannot however be safely cut off.

Boundary conditions on any flow are applied by having walls made of atoms. These atoms should vibrate with thermal energy at the same temperature as the interior, or else they will be a heat source or sink. For some phenomena, it may be necessary to have a wall of several layers of atoms.

The Newtonian dynamics of the moving atoms is Hamiltonian. A symplectic integrator, see §7.9, can therefore be used to preserve volumes in the $pq$-phase-space of trajectories of solutions. A symplectic integrator commonly used in molecular dynamics is the Störmer–Verlet algorithm

$$\dot{\mathbf{r}}^{n+\frac{1}{2}} = \dot{\mathbf{r}}^{n} + \frac{1}{2m}\Delta t\, \mathbf{F}(\mathbf{r}^{n}),$$

$$\mathbf{r}^{n+1} = \mathbf{r}^{n} + \Delta t\, \dot{\mathbf{r}}^{n+\frac{1}{2}},$$

$$\dot{\mathbf{r}}^{n+1} = \dot{\mathbf{r}}^{n+\frac{1}{2}} + \frac{1}{2m}\Delta t\, \mathbf{F}(\mathbf{r}^{n+1}),$$

where $m$ is the mass and $\mathbf{F}(\mathbf{r})$ the force.

Molecular dynamics computations last many time-steps. Energy can be lost slowly over a long time. The loss of energy is a cooling of the thermal motion of the atoms. Some process of maintaining the long-term temperature is required. It is possible to rescale the velocities of all the atoms to restore the temperature. Alternatively the imposed temperature of the atoms vibrating in the walls can be adjusted to inject heat into the interior.

Molecules can be constructed by binding several atoms with strong bonds, i.e. setting the interaction energy $\epsilon$ in the Lennard–Jones potential for the bonds to be many $kT$. The problem with strong bonds is that the associated vibration frequency is very high, requiring very small time-steps. If the vibration modes are of no interest, the stiff bonds can be replaced by rigid bonds, in which tensions are applied in the bonds to constrain the bond length to a particular value. There are a number of algorithms, such as SHAKE and RATTLE, which maintain the constraints. It is a sparse linear problem to find the values of the tensions which make the velocities instantaneously orthogonal to the constraints. It is unfortunately a nonlinear problem to make small adjustments to those values which guarantee that the constraints are satisfied exactly at the end of the finite time-step. Making the bonds rigid freezes out degrees of freedom and so reduces the specific heat of the system. Freezing bonds introduces other more subtle and possibly undesirable changes, such as making some included angles of a freely hinged trimer more probable than others.

Macroscopic observables and macroscopic properties are obtained from a molecular dynamics simulation by taking averages; averages over all the atoms if the system is spatially homogeneous, averages over time if the time exceeds that required by any physical process to evolve, and averages over different realisations. Averages over a finite system with $N$ independent atoms will have statistical fluctuations $O(N^{-1/2})$. This uncertainty can be unfortunate.

The numerical algorithms for molecular dynamics are these days sophisticated. An amateur should not attempt to code them. There exist many packages, a good number free, and a good number parallelised for GPU devices.

## 16.2 Lattice Gas

The Lattice Gas model greatly simplifies the motion of molecules in a gas. The molecules move only along a lattice, starting and ending each time-step at a node of the lattice. In one time-step, the molecules move singly to an adjacent node. Thus the positions and velocities are discretised. The state of the gas is therefore described by a vector of Boolean variables for each node. The variables say yes or no that there is one particle at the node travelling in the particular direction of an adjacent node. Sometimes there is an extra possibility of one rest particle at the node which does not move at the time-step.

The interaction forces between the molecules are replaced by simple rules about the collisions between molecules arriving at a node from different directions. Particles moving in opposite directions along the connection between two nodes pass through one another, because collisions only occur at nodes. Collisions at a node must conserve mass and momentum. If two particles approach head on at a node, they exit along opposite but new directions in order for there to be a genuine collision. Sometimes there are several possible outcomes which equally conserve mass and momentum, see Figure 16.1, in which case one outcome is selected at random. If the number of rest particles changes, then energy is not conserved. As the collisions are local and transport

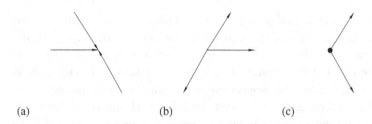

(a)                              (b)                              (c)

Figure 16.1 A three-particle collision in a Lattice Gas on a hexagonal lattice.
(a) Three particles approach a node, (b) two change direction or (c) two change
direction and one becomes stationary.

is only between adjacent nodes and is unconditional, the Lattice Gas dynamics
is highly parallelisable. Cellular autonoma have been built to implement the
dynamics.

A nonslip boundary condition is applied by reversing the direction of any
particle arriving at the boundary. A no-flux slip boundary condition is applied
by reversing the normal component of the velocity of any particle arriving at
the boundary. Flows are driven by inflow over a boundary.

A Lattice Gas behaves like a compressible gas. The distance between adjacent nodes, $\Delta x$, is the mean-free-path. The speed at which the particles move
in one time-step $\Delta t$ between adjacent nodes, $c = \Delta x/\Delta t$, is the speed of sound
in the compressible gas. Because the collision rules have greatly simplified the
interactions between the particles, the physics of the Lattice Gas at the smallest microscale is not the correct physics of any real gas. However the physics
at the large macroscale is that of a continuum compressible gas. The density
of the gas $\rho$ is found by averaging the number of particles at a node coming
from all the different directions and multiplying by the mass of a particle per
elementary lattice volume. The macroscopic velocity $\mathbf{u}$ is found by averaging
the sum of the velocities of the different particles at a node. The averaging
can be a time average at a single node, a space average over nearby nodes
or an ensemble average over different realisations. These averages have large
$O(N^{-1/2})$ fluctuations, where there are $N$ independent members of the sample
being averaged.

Because all the particles move on the lattice at the speed of sound, the velocity of the gas is limited to be subsonic. The dynamics is not realistic for Mach
numbers close to unity. The dynamics is also not Galilean invariant at such
speeds. There are also problems simulating low Mach number flows, nearly
incompressible flows, because the $O(cN^{-1/2})$ fluctuations can swamp a mean
velocity which is much less than $c$.

The macroscopic pressure is found as the average of the sum of the square of
the velocities of the particles at a node, multiplied by the mass of a particle per

elementary lattice volume, so $p = \rho c^2$. This is a particularly simple equation of state coming from the simplified collision rules.

For slowly varying in time and large-scale features, the Lattice Gas behaves like a continuum compressible gas. The collisions between the particles lead to a diffusion of momentum, i.e a viscous behaviour. Viscosity is the linear relation between the second-order stress tensor and the second-order velocity gradient tensor, and so in general is a fourth-order tensor. For a Lattice Gas, this fourth-order tensor is proportional to a sum over all the directions $\Delta\mathbf{x}$ at a node $\sum \Delta\mathbf{x}\,\Delta\mathbf{x}\,\Delta\mathbf{x}\,\Delta\mathbf{x}$. For a Cartesian square lattice, this fourth-order tensor is not isotropic; in fact the only components of stress are $\sigma_{xx} = 2\mu\partial u/\partial x$ and $\sigma_{yy} = 2\mu\partial v/\partial y$, i.e. there are no shear stresses. In order to have the normal behaviour with an isotropic viscosity, it is necessary to use an hexagonal lattice in two dimensions. There are issues in three dimensions, where it is necessary to use diagonal connections, which require particles to travel faster. Note that the kinematic viscosity $v \propto c\Delta x = \Delta x^2/\Delta t$, with a numerical factor which depends on the density. These two expressions for the viscosity show that the numerical algorithm will automatically satisfy the CFL and the diffusion stability criteria of not advecting or diffusing more than one lattice space in one time-step.

Being a compressible gas, the Lattice Gas algorithm does not have a Poisson problem to solve for the pressure. However it is not immediately clear that this means the algorithm is faster than a simple finite difference approach. Using $v = c\Delta x$, the Reynolds number $Re = UL/v = (U/c)(L/\Delta x)$. This means that in order to have moderate Reynolds numbers, say $Re = 10$, and only a slightly compressible gas, say $U/c = 0.1$, one needs a large domain, say $L = 10^2\Delta x$. For the flow to evolve at velocity $U$ in such a domain, one needs to compute to time $L/U = 10^3\Delta t$, i.e. for many small time-steps. Moreover to make the statistical fluctuations in the velocity smaller than 10% of $0.1c$ one has to average over $10^4$ realisations. The advantage of not solving a Poisson problem is outweighed by the small time-steps for sound propagation and the unavoidable statistical fluctuations.

## 16.3 Lattice Boltzmann

The Lattice Boltzmann method removes the statistical fluctuations of the Lattice Gas by replacing the on-off Boolean variables for the existence of a single particle by a continuously varying population of particles travelling in a direction at a node. The Boltzmann description of a gas is a continuum description by a probability distribution $f(\mathbf{v}, \mathbf{x}, t)$ for a particle to have velocity $\mathbf{v}$ at location $\mathbf{x}$ at time $t$. From the distribution one can evaluate the density $\rho$ and the mean velocity **u** of the gas

$$\rho(\mathbf{x}, t) = m \int f(\mathbf{v}, \mathbf{x}, t) \, d^3v, \qquad \rho\mathbf{u}(\mathbf{x}, t) = m \int \mathbf{v} f(\mathbf{v}, \mathbf{x}, t) \, d^3v,$$

where $m$ is the mass of a particle. The probability distribution is governed by the Boltzmann equation

$$\frac{\partial f}{\partial t} + \mathbf{v} \cdot \frac{\partial f}{\partial \mathbf{x}} + \mathbf{a} \cdot \frac{\partial f}{\partial \mathbf{v}} = C\{f, f\},$$

where **a** is the acceleration due to any forces and $C\{f, f\}$ is a collision operator. When there are no forces, the distribution relaxes towards a thermodynamic equilibrium given by the Maxwellian distribution about the mean velocity **u**

$$f^{\text{eq}}(\mathbf{v}, \mathbf{x}, t) = \frac{\rho(\mathbf{x}, t)/m}{(2\pi kT/m)^{3/2}} e^{-\frac{1}{2}m(\mathbf{v}-\mathbf{u})^2/kT}.$$

For a mean velocity **u** small compared with the thermal velocities $(kT/m)^{1/2}$, i.e. for a low Mach number flow, the exponential can be expanded to second order

$$e^{-\frac{1}{2}m(\mathbf{v}-\mathbf{u})^2/kT} = e^{-\frac{1}{2}m\mathbf{v}^2/kT} \left(1 + \frac{m}{kT}\mathbf{u} \cdot \mathbf{v} + \frac{m}{2kT}\left(\frac{m}{kT}(\mathbf{u} \cdot \mathbf{v})^2 - u^2\right) + \cdots\right).$$

A major simplification of the collision term is to make the BGK (Bhatnager–Gross–Krook 1954) approximation that the distribution relaxes exponentially towards the thermodynamic equilibrium on a time-scale $\tau$, i.e.

$$\frac{\partial f}{\partial t} + \mathbf{v} \cdot \frac{\partial f}{\partial \mathbf{x}} = -\frac{1}{\tau}(f - f^{\text{eq}}).$$

The Lattice Boltzmann method is a discretised version of this equation above. The location of the particles is restricted to lattice nodes $\mathbf{x}_n$ and the velocities are restricted to values $\mathbf{v}_i$ which enable the particles to move between adjacent nodes in a single time-step $\Delta t$. One writes $f_i(\mathbf{x}_n, t)$ for $f(\mathbf{v}_i, \mathbf{x}_n, t)$. Cartesian lattices are used with some additional diagonal links between adjacent nodes. This means that some particles have to travel faster than others, $\sqrt{2}$ and $\sqrt{3}$

faster. Lattices used in two and three dimensions are the following.

| *D2Q9*: | Directions | Speeds | Weights $w_i$ |
|---|---|---|---|
| | $(0,0)$ | $0$ | $\frac{4}{9}$ |
| | $(\pm 1, 0), (0, \pm 1)$ | $c$ | $\frac{1}{9}$ |
| | $(\pm 1, \pm 1)$ | $\sqrt{2}c$ | $\frac{1}{36}$ |
| *D3Q27*: | | | |
| | $(0, 0, 0)$ | $0$ | $\frac{8}{27}$ |
| | $(\pm 1, 0, 0), (0, \pm 1, 0), (0, 0, \pm 1)$ | $c$ | $\frac{2}{27}$ |
| | $(0, \pm 1, \pm 1), (\pm 1, 0, \pm 1), (\pm 1, \pm 1, 0)$ | $\sqrt{2}c$ | $\frac{1}{54}$ |
| | $(\pm 1, \pm 1, \pm 1)$ | $\sqrt{3}c$ | $\frac{1}{216}$. |

In three dimensions there is also *D3Q15* which misses out the 12 points in the middle of the faces, $(\pm 1, \pm 1, 0)$ etc., and has weights $\frac{2}{9}, \frac{1}{9}, \frac{1}{72}$. The weights multiplied by the number of particles with those weights sums to unity. The weights decrease corresponding to the Maxwell distribution on the speeds, $e^{-m(\beta c)^2/2kT}$ with $\beta = 0, 1, \sqrt{2}$ and $\sqrt{3}$.

The algorithm uses a split time-step. In the first part the populations are advected, while in the second part they are relaxed towards their thermodynamic equilibrium. Thus for the first step

$$f_i^*(\mathbf{x}_n, t) = f_i(\mathbf{x}_n - \mathbf{v}_i \Delta t, t - \Delta t),$$

where the time-step $\Delta t$ is set as the time to move between lattice positions, i.e. $\Delta x = c\Delta t$. From these advected populations, one calculates the new local density and mean velocity

$$\rho(\mathbf{x}_n, t) = \sum_i f_i^*(\mathbf{x}_n, t), \quad \rho(\mathbf{x}_n, t)\mathbf{u}(\mathbf{x}_n, t) = \sum_i \mathbf{v}_i f_i^*(\mathbf{x}_n, t).$$

Choosing a scaling $\frac{1}{2}mc^2 = \frac{3}{2}kT$, the new thermodynamic equilibrium populations are

$$f_i^{\text{eq}}(\mathbf{x}_n, t) = \rho w_i \left( 1 + 3\frac{\mathbf{u} \cdot \mathbf{v}_i}{c^2} + \frac{9}{2}\frac{(\mathbf{u} \cdot \mathbf{v}_i)^2}{c^4} - \frac{3}{2}\frac{u^2}{c^2} \right),$$

with weights $w_i$ for the different directions given by the table above. The second part of the split time-step for the relaxation is

$$f_i(\mathbf{x}_n, t) = f_i^*(\mathbf{x}_n, t) - \frac{\Delta t}{\tau} \left( f_i^{\text{eq}}(\mathbf{x}_n, t) - f_i^*(\mathbf{x}_n, t) \right).$$

For numerical stability one needs $\Delta t < 2\tau$, i.e. the thermodynamic relaxation must be given more than half a time-step, preferably several.

The pressure in the Lattice Boltzmann Gas is

$$p(\mathbf{x}_n, t) = \sum_i f_i(\mathbf{x}_n, t)\mathbf{v}_i \cdot \mathbf{v}_i = \begin{cases} \frac{2}{3}\rho c^2 & \text{for } D2Q9, \\ \rho c^2 & \text{for } D3Q27. \end{cases}$$

Thus the speed of sound is constant, respectively $\sqrt{\frac{2}{3}}c$ and $c$ in 2D and 3D. For slowly varying and large-scale features, the Lattice Boltzmann Gas behaves like a compressible continuum gas with a kinematic viscosity

$$(\tau - \tfrac{1}{2}\Delta t)c^2.$$

To simulate low viscosity fluids, the relaxation time is often set only marginally about the stability limit $\tau = \frac{1}{2}\Delta t$.

Being a compressible gas, the pressure is a local variable and not found by solving a time-consuming Poisson problem. That advantage, however, has to be set against the very small time-step, that needed to propagate sound one lattice spacing, $\Delta t = \Delta x/c$. As one has made a low Mach number approximation in the expansion of the Maxwellian distributions, the speed of sound should be irrelevant to the physics, yet it is controlling the algorithm. There is a further disadvantage of the method, that to find the three components of velocity in 3D and pressure, four unknowns, one has to compute 27 populations $f_i$ in $D3Q27$, and in two dimensions nine populations in $D2Q9$ compared with two velocity components plus pressure.

There exist packages.

## 16.4 Dissipative particle dynamics

Dissipative particle dynamics is a mesoscale approach, midway between molecular dynamics and continuum fluid dynamics. Like molecular dynamics it has particles moving in all possible directions at all possible speeds, i.e. the particles are not restricted to move on a lattice as in Lattice Gases and the Lattice Boltzmann method. The interactions between the particles are, however, very different from the interactions between molecules.

The particles are spheres. They interact in a pairwise additive way with nearby neighbours within some cutoff distance, typically a particle diameter or two. The interaction force for a pair is only in the direction of the line joining their centres, i.e. there is no tangential force. There are three parts to the interaction force. An elastic part depends on the separation of the particles and is a soft repulsion to reduce the particles overlapping. A friction part is resistive and proportional to the rate at which the particles separate or approach. This friction part dissipates energy, so that it is necessary to inject energy through a Brownian force. The magnitude of this Brownian force is related to the size of the time-step and the friction constant by a fluctuation-dissipation theorem, see §16.5.2.

As a mesoscale approach, dissipative particle dynamics is not good at representing microscale molecular dynamics or macroscale fluid dynamics. It is impossible to relate the parameters of dissipative particle dynamics, the spring constant and in particular the friction constant, to the properties of molecules and their interactions. There is a serious problem that the diffusion of particles is as efficient as the diffusion of momentum (low Schmidt number) which is not appropriate to real liquids. For fluid dynamics calculations, there is no obvious way to improve the spatial resolution of the flow while keeping the Reynolds number to a set value. I see no useful role for this approach.

## 16.5 Stokesian dynamics

### 16.5.1 Hydrodynamic interactions

Stokesian dynamics was developed by Brady and Bossis in 1988 to study the motion of many particles in a suspension. One might be interested in the sedimentation or the Brownian diffusion of the particles, or the rheology of the suspension when subjected to a bulk shearing. The particles must be small and the suspending fluid viscous so that the Reynolds number of the flow between the particles is small, and so the inertia of the particles and the fluid is negligible. Further, the particles must be rigid and spheres. More complex particles can be built by gluing together several rigid spheres.

Just the hydrodynamic forces on the moving particles are calculated, not the details of the flow between the particles. Near-field interactions between nearby particles are calculated well by lubrication forces. Far-field interactions, which are strictly multiparticle interactions, are calculated only to a certain level of approximation, which could be improved but rarely is.

At low Reynolds numbers the forces on the particles are linear in their velocities. One therefore seeks the drag law with a resistance tensor $\mathbf{R}$

$$\mathbf{F} = \mathbf{R}(\mathbf{U}^\infty - \mathbf{U}),$$

for the forces $\mathbf{F}$ when the particles move at velocity $\mathbf{U}$ in an externally imposed flow $\mathbf{U}^\infty(\mathbf{x})$. The velocity $\mathbf{U}$ includes the translational velocity and the rotational velocity of each of the $N$ particles, i.e. a $6N$ component vector. Similarly the force $\mathbf{F}$ includes the net force on each particle, as well as the couple, again a $6N$ component vector. In fact one needs to add a few more components: to the velocity one adds the externally imposed strain-rate $\mathbf{E}^\infty$, and to the force one adds the stresslet on each particle, the stresslet being the symmetric force-dipole to go with the antisymmetric force-dipole of the couple. The inverse of the resistance tensor $\mathbf{R}$ is the mobility tensor $\mathbf{M}$ which gives the velocity driven by a force,

$$\mathbf{U} = \mathbf{U}^\infty + \mathbf{MF}.$$

Stokesian dynamics calculates the resistance tensor $\mathbf{R}$ by adding the separate contributions from the far-field and the near-field, and then removing duplicates, so

$$\mathbf{R} = (\mathbf{M}_{ff})^{-1} + \mathbf{R}_{nf}.$$

The multiparticle interactions in the far-field are handled in the mobility tensor $\mathbf{M}_{ff}$ where one can add linearly the contributions to the velocity from the force multipoles from all the particles. A Faxen correction is applied to the velocities of the particles to account for the quadratic variation of the induced flow across the diameter of a particle. The standard Stokesian dynamics uses only the force and force-dipoles, so that the first ignored multipole is a quadrupole. That ignored quadrupole only contributes at $O(1/r^6)$ in sedimentation and $O(1/r^7)$ in rheology, where $r$ is an interparticle separation.

While one can add linearly the independent contributions from the different particles to the far-field mobility tensor $\mathbf{M}_{ff}$, its inverse $\mathbf{M}_{ff}^{-1}$ has many multiparticle interactions. For example, the force on particle 1 due to particle 2 moving has a term $O(1/(r_{23}r_{31}))$ involving a third particle, where $r_{ij}$ is the separation between particles $i$ and $j$.

The near-field contribution to the resistance tensor is a pairwise sum of the exact two-particle interactions $\mathbf{R}_2$ with far-field parts in $\mathbf{M}_{ff}$ removed.

$$\mathbf{R}_{nf} = \mathbf{R}_2 - \mathbf{R}_{2ff}.$$

The pairwise sum ignores multiparticle contributions, which is a reasonable approximation if one thinks that the near-field is just for lubrication forces where a third particle cannot intervene in a narrow separation between two particles. However in practice the near-field is often applied out to two particle diameters.

Note that one eventually needs an approximation to the mobility tensor in order to find the velocities at which the particles move. A mobility tensor constructed from a pairwise sum of the exact two-particle interactions to a mobility tensor, however, leads to remote particles inducing velocities which cause particles to overlap.

For $N$ particles, the cost of constructing $\mathbf{M}_{\text{ff}}$ in $O(N^2)$ operations, and the cost of inverting it to form $\mathbf{M}_{\text{ff}}^{-1}$ is $O(N^3)$ operations. As the far-field evolves slowly, this inversion need not be performed every time-step. The near-field contribution $\mathbf{R}_{\text{nf}}$ involves only a few neighbours of each particle and so requires $O(N)$ operations. The major computational cost is the $O(N^3)$ inversion of the full resistance tensor $\mathbf{R}$ in order to find the velocity at which the particles move in response to any applied forces or externally applied straining flow. There are possibilities to accelerate this inversion, say by conjugate gradients, or by iterating based on recognising the dominant contributions from the nearby particles.

### 16.5.2 Brownian motion

Small particles in a suspension, in particular in colloidal dispersions, are subject to Brownian motion. With Brownian motion, the positions of the particles are described by a probability distribution function $P(\mathbf{x}, t)$ over the possible positions. For simple systems of a couple of particles, the diffusive motion is best studied numerically by solving a Fokker–Plank or Smolochowski advection-diffusion partial differential equation

$$\frac{\partial P}{\partial t} + \nabla \cdot (\mathbf{U} P) = \nabla \cdot \mathbf{D} \cdot \nabla P.$$

Here the diffusivity related to the mobility by the Stokes–Einstein expression

$$\mathbf{D} = kT\mathbf{R}^{-1},$$

with $kT$ the Boltzmann temperature.

For more complex systems with many degrees of freedom, a numerical simulation of random walks of the diffusive Brownian motion through many different configurations can be more economical. The computational trade-off is between the rapidly increasing storage requirements for the partial differential equation over many dimensions against the slowly decreasing $N^{-1/2}$ statistical errors when averages are made over $N$ simulated random walks.

To simulate the random walk, one adds at each time-step a random displacement $\boldsymbol{\xi}$ to the systematic motion driven by a steady force $\mathbf{F}^S$,

$$\mathbf{x}^{n+1} = \mathbf{x}^n + \mathbf{R}^{-1}\mathbf{F}^S \Delta t + \boldsymbol{\xi}^n.$$

The random displacements at different time-steps should be uncorrelated. Be warned some computer random number generators generate a sequence of numbers with a high correlation between adjacent numbers. The displacements should have zero mean, $\langle \boldsymbol{\xi}^n \rangle = 0$, and a variance to give the correct dispersion of the random walk, $\langle \boldsymbol{\xi}^n \boldsymbol{\xi}^n \rangle = 2\mathbf{D}\Delta t$. The random displacements are generated by

$$\boldsymbol{\xi}^n = (2\Delta t)^{1/2}\mathbf{A}\mathbf{r}^n,$$

with amplitude $\mathbf{A}$ and a unit random vector $\mathbf{r}^n$ with

$$\langle \mathbf{r}^n \rangle = 0, \quad \text{and} \quad \langle \mathbf{r}^n \mathbf{r}^n \rangle = \mathbf{I}.$$

Each of the components of $\mathbf{r}^n$ can be independently generated with zero mean and unit variance. The amplitude $\mathbf{A}$ is a sort of square root of the diffusivity,

$$\mathbf{A}\mathbf{A}^T = \mathbf{D},$$

which can be found using a Cholesky decomposition of $\mathbf{D}$, see §8.2.2. With displacements proportional to $(\Delta t)^{1/2}$, quite small time-steps are often needed when there is fine spatial structure to be resolved.

The random displacements in the time-step can be thought of as the result of a random Brownian force $\mathbf{F}^{\text{BM},n}$, i.e.

$$\boldsymbol{\xi}^n = \mathbf{R}^{-1}\mathbf{F}^{\text{BM},n}\Delta t.$$

The Brownian force during one time-step is uncorrelated with the Brownian force at other times. It has a size $O((\Delta t)^{-1/2})$ given by a discretized version of the *fluctuation dissipation* theorem

$$\langle \mathbf{F}^{\text{BM},m}\mathbf{F}^{\text{BM},n} \rangle = 2kT\mathbf{R}\delta_{mn}/\Delta t.$$

The random force can be generated like $\boldsymbol{\xi}^n$ from a unit random vector $\mathbf{r}^n$ by

$$\mathbf{F}^{\text{BM},n} = \mathbf{B}\mathbf{r}^n, \quad \text{where} \quad \mathbf{B}\mathbf{B}^T = 2kT\mathbf{R}/\Delta t.$$

Here $\mathbf{B}$, a sort of square root, is related to the earlier square root of the diffusivity as $\mathbf{B} = \mathbf{R}\mathbf{A}(2/\Delta t)^{1/2}$.

In normal Stokesian dynamics, the inertia of the fluid and the particles is negligibly small. One can include just the particle inertia in a *Langevin equation* description with a mass/inertia tensor $\mathbf{m}$

$$\mathbf{m}\ddot{\mathbf{x}} = \mathbf{R}\,(\mathbf{U}^\infty - \dot{\mathbf{x}}) + \mathbf{F}^S + \mathbf{F}^{\text{BM}}.$$

The random Brownian forces $\mathbf{F}^{\text{BM}}$ are given by the same fluctuation-dissipation theorem. The Langevin equation gives the same diffusion behaviour at long times, and has a fast new process of velocity relaxation at short times. If one is only interested in the diffusional behaviour, it is expensive numerically to resolve the uninteresting fast short-time process.

## Drift term

There is a problem with the naive algorithm above when the diffusivity changes with the configuration of the particles. Unless a correction is made, configurations with a high diffusivity are erroneously depleted as a result of vigorous random walks away from such configurations compared with more feeble walks returning from surrounding configurations. The required correction found by Ermak and McCammon in 1978 is to add a drift term

$$\mathbf{x}^{n+1} = \mathbf{x}^n + \mathbf{R}^{-1}\mathbf{F}^S \Delta t + \boldsymbol{\xi}^n + \nabla\cdot\mathbf{D}\Delta t.$$

Calculating the divergence of the diffusivity is not simple. Fortunately it can be avoided with a midpoint scheme, as suggested by Fixman in 1978.

$$\mathbf{x}^* = \mathbf{x}^n + \tfrac{1}{2}\mathbf{R}^{-1}(\mathbf{x}^n)\left(\mathbf{F}^S(\mathbf{x}^n) + \mathbf{F}^{\mathrm{BM},n}\right)\Delta t,$$
$$\mathbf{x}^{n+1} = \mathbf{x}^n + \mathbf{R}^{-1}(\mathbf{x}^*)\left(\mathbf{F}^S(\mathbf{x}^*) + \mathbf{F}^{\mathrm{BM},n}\right)\Delta t.$$

During the two parts of the time-step, the Brownian force is constant while the mobility changes.

## Rheology

In studying the rheology of a suspension, one often has to evaluate the moment of the forces $\langle\mathbf{xF}\rangle$. There is a problem with large $O((\Delta t)^{-1/2})$ size of the fluctuating random Brownian forces. One has to average over very many, $O(1/\Delta t)$, realisations before the statistical error in the average reduces to being $O(1)$, and even more to make the statistical errors suitably small. Now in the time-step from $\mathbf{x}^n$ to $\mathbf{x}^{n+1}$, the Brownian force $\mathbf{F}^{\mathrm{BM}}$ is uncorrelated with the initial position $\mathbf{x}^n$. The force just is correlated with the displacement $\mathbf{x}^{n+1}-\mathbf{x}^n$. Considering the variation of the position during the time-step to vary linearly in time, the correlated contribution to the moment of the force is just $\tfrac{1}{2}(\mathbf{x}^{n+1} - \mathbf{x}^n)\mathbf{F}^{\mathrm{BM}}$. This is the product of a term proportional to $O((\Delta t)^{1/2})$ and a term proportional to $O((\Delta t)^{-1/2})$, i.e. an $O(1)$ quantity, which requires far fewer realisations to obtain a significant average.

# 16.6 Force Coupling Method

The Force Coupling Method (FCM) of Maxey in 2001 is an alternative to Stokesian dynamics for studying the flow of suspensions of particles. Whereas the Reynolds number for the flow between the particles must be small in Stokesian dynamics, it can be moderate in the FCM. Whereas the flow is not explicitly calculated in Stokesian dynamics, it is fully found in the FCM.

The Force Coupling Method represents each particle $\alpha$ by a distributed force monopole of strength $\mathbf{F}^\alpha$ and distributed force dipole of strength $\mathbf{Q}^\alpha$. The Navier–Stokes equation for the flow within the suspension is then solved with these distributed forces

$$\mathbf{f}(\mathbf{x}, t) = \sum_\alpha \mathbf{F}^\alpha \Delta_M (\mathbf{x} - \mathbf{X}^\alpha) + \sum_\alpha \mathbf{Q}^\alpha \cdot \nabla \Delta_D (\mathbf{x} - \mathbf{X}^\alpha).$$

The form of the distributions of force is normally taken to be a Gaussian $\Delta(\mathbf{x}) = (2\pi\sigma^2)^{-3/2} \exp(-\mathbf{x}^2/2\sigma^2)$, and this Gaussian is useful in pseudospectral methods of solving the Navier–Stokes equation. For a sphere of radius $a$, the monopole size is set to $\sigma_M = a(\pi)^{-1/2}$, and the dipole size is set to $\sigma_D = a(36\pi)^{-1/6}$. For finite differences, the spatial resolution needs $\Delta x \lesssim a/3$.

The magnitude of the force $\mathbf{F}^\alpha$ that the particle exerts on the fluid comes from the equation of motion of the particle

$$\mathbf{F}^\alpha = \mathbf{F}^{\alpha,\text{ext}} + (m_F - m_P)\frac{d\mathbf{V}^\alpha}{dt},$$

where $\mathbf{F}^{\text{ext}}$ is the external force on the particle, and $m_P$ and $m_F$ are the masses of the particle and of the fluid it displaces. The velocity of the particle $\mathbf{V}^\alpha$ is taken as the average of the fluid velocity within the range of the distributed monopole

$$\mathbf{V}^\alpha(\mathbf{x}, t) = \int \mathbf{u}(\mathbf{x}, t) \, \Delta_M (\mathbf{x} - \mathbf{X}^\alpha) \, dV(\mathbf{x}).$$

The magnitude of the antisymmetric part of force dipole $\mathbf{Q}^\alpha$ is found in a similar way considering the torque and angular velocity of the particle in place of the force and the translational velocity. The magnitude of the symmetric part of the force dipole, the particle stresslet, is adjusted so that the average strain-rate within the particle vanishes, i.e.

$$0 = \int \left(\nabla\mathbf{u} + \nabla\mathbf{u}^T\right)(\mathbf{x}, t) \, \Delta_D (\mathbf{x} - \mathbf{X}^\alpha) \, dV(\mathbf{x}).$$

The strain-rate should vanish everywhere within a rigid particle. Requiring that the average strain-rate vanishes is an acceptable approximation while the particles are well separated. When particles become close in a flow, the above Force Coupling Method gives considerable localised straining in the regions of closest approach. i.e. the particles are rather squashy. For these situations, lubrication forces and torques can be added to the external forces and torques in the equations of motion of the particles.

The Force Coupling Method represents each particle by just a force monopole and a force dipole. Higher multipoles could be added. An alternative is to place a number of force monopoles around the boundary of each particle. The magnitudes of these monopoles are adjusted to make the strain-rate vanish at several places within the particle. This approach is called the *immersed boundary method* of Peskin (1972) where the adjustment of the magnitudes is called *penalisation*.

## 16.7 Granular media simulations

There was no good continuum description of the flow of a dense granular medium until the recent development of the $\mu(I)$-rheology. Before that it was necessary to simulate the motion of many individual grains in order to predict a macroscale behaviour. While the motion of $O(10^5)$ grains could be followed, that is far fewer grains than in a practical application. Fortunately most large-scale features of the flow are independent of the number of grains.

Rather similar to molecular dynamics, the rigid grains move according to Newton's equations of motion. Different from molecular dynamics, both the rotation and translation of the grains are computed. The grains interact through contact forces with those neighbours that they are touching. The numerical approach is often called the Discrete Element Method (DEM).

The first question is what is the shape of the particles to be used. Real grains have an irregular shape. Most simulations use spherical grains, occasionally polygons. For spheres, it is easy to determine whether two grains are in contact. For other shapes the calculation is more complex.

The contact forces between grains have a normal and a tangential component. The normal component resists touching grains becoming closer, and vanishes if the grains are trying to separate, i.e. there are no cohesive forces between dry grains. In the Contact Dynamics approach of Moreau and Jean (1994), the constraints that the normal component of the relative acceleration of touching grains must be nonnegative lead to a sparse system of linear equations for the normal forces, which themselves must be nonnegative. Deciding which contacts are active in order to satisfy the two nonnegative conditions is an awkward search. An alternative approach is to introduce an elastic spring for the normal forces. Typical grains are very rigid and deform little. To avoid resolving very small movements, most simulations make the grains much softer. So long as the deformations are small compared with the size of the grains, the artificial softness should be unimportant. Some simulations add a frictional normal force proportional to the normal relative velocity. This is unrealistic and unnecessary, because dissipation is amply supplied by the tangential forces.

For dry grains, the tangential component $F$ of the contact forces should be that of Coulombic friction. Thus there should be no relative slip between the touching surfaces if the tangential force is less than $\mu$ times the normal force $N$, $F < \mu N$, and if the surfaces are slipping past one another the force should be resistive and $\mu$ times the normal force, $F = \mu N$. Calculating these friction forces is more complicated than calculating the normal forces with their non-negative constraints, because the friction forces for one configuration are not unique, but depend on the history of how the configuration came into being. One can make the tangential forces well and easily determined by adopting an elastic-perfectly-plastic model. For small relative tangential displacements, there is a linear elastic spring law with a spring constant $k$. As the displacement increases, the tangential force is limited by the Coulomb value $\mu N$, and an elastic displacement set to the limiting value of $\mu N/k$. All displacements beyond this value are plastic. When the displacements begin to decrease, all the plastic part is forgotten and the elastic part is decreased. The results of simulations tend to depend weakly on the value of the intergrain coefficient of friction $\mu$, because quite small values are required to arrest tangential slip early during a contact, and once arrested the value of $\mu$ is mostly irrelevant. It should be noted that just three touching grains frustrate counterrotating pairs.

There can be problems with perfectly flat boundaries and spherical grains, because the grains can become isolated and roll away. It is better to have rough boundaries made of fixed grains.

As well as boundary conditions, it is important that the initial conditions are specified clearly, because the tangential forces depend on past history. There are various approaches to setting up the initial configuration. One is to add grains one at a time at the top boundary and allow them to fall into contact with the grains below, with the possibility of induced rearrangements.

The numerical method to be used to integrate Newton's equations of motion is as in molecular dynamics the Verlet algorithm, see §7.9, although a slow numerical loss of energy is not an issue when the physics has a dissipation mechanism. The loss of energy through dissipation allows a pile of sand to not be flattened by gravity as a fluid is.

There exist packages.

## 16.8 Smooth Particle Hydrodynamics

Smooth Particle Hydrodynamics is an approach for inviscid compressible gas flows. It is a meshless approach, in that there is no finite difference grid and there is no finite element triangulation of the domain. Instead the volume of the gas is allocated to fluid particles. Each particle has a mass $m_i$ which does not change in time. The particle has a density $\rho_i(t)$ which does vary across the flow and varies in time. From the density and the mass one finds the volume of the particle as $m_i/\rho_i$. Also from the density and an equation of state one finds the pressure $p_i(t)$ of the fluid particle

$$p_i(t) = p_i(0) \left( \frac{\rho_i(t)}{\rho_i(0)} \right)^{\gamma} .$$

The fluid particles have position $\mathbf{r}_i(t)$ and velocity $\mathbf{v}_i(t) = \dot{\mathbf{r}}_i$. They have no shape and no internal structure. They are not linked to their neighbours as in finite differences and finite elements.

The dynamics is governed by two equations. The density changes due to a divergence in the velocity field

$$\frac{d\rho_i}{dt} = -\rho_i \, (\boldsymbol{\nabla} \cdot \mathbf{u})_i \, ,$$

and the particle velocity changes due to the pressure gradient

$$\rho_i \frac{d\mathbf{v}_i}{dt} = - \, (\boldsymbol{\nabla} p)_i \, .$$

These two equations for the evolution of the flow need spatial derivatives of the velocity and pressure whose spot values are known at the location of the particles.

The key idea of Smooth Particle Hydrodynamics is to construct the full velocity and pressure fields from the spot values at the particles by using a smoothing function. For the pressure $p(\mathbf{x})$, one can form a smoothed version $p^{\text{sm}}(\mathbf{x})$ as a weighted integral of nearby values

$$p^{\text{sm}}(\mathbf{x}) = \int p(\mathbf{x}')w((\mathbf{x}' - \mathbf{x})/h) \, dV(\mathbf{x}').$$

While a Gaussian could be used for the weighting kernel $w$, a function which vanishes outside a finite range such as the cubic $B$-spline of §11.1.1 is better, e.g. in three dimensions

$$w(r) = \frac{1}{2\pi} \begin{cases} 1 - \frac{3}{2}r^2 + \frac{3}{4}r^3 & 0 \le r \le 1, \\ \frac{1}{4}(2 - r)^3 & 1 \le r \le 2, \\ 0 & r \ge 2, \end{cases}$$

with normalisation constant $1/2\pi$ replaced by $10/7\pi$ in two dimensions. The smoothed function differs from the original function by $O(h^2)$. A numerical approximation $p^{\text{num}}(\mathbf{x})$ is formed by replacing the integral with a sum over the spot particle values

$$p^{\mathrm{num}}(\mathbf{x}) = \sum_j \frac{m_j}{\rho_j} p_j \, w((\mathbf{x}_j - \mathbf{x})/h).$$

This approximation suffers large $O(N^{-1/2})$ statistical fluctuations where $N$ is the number of particles within the sampling range. To make this error small requires very large numbers of particles in the simulation. Fortunately these statistical fluctuations can be eliminated by renormalising the sum

$$p^{\mathrm{num2}}(\mathbf{x}) = \frac{1}{C} \sum_j \frac{m_j}{\rho_j} p_j \, w((\mathbf{x}_j - \mathbf{x})/h)$$

$$\text{where} \quad C = \sum_j \frac{m_j}{\rho_j} w((\mathbf{x}_j - \mathbf{x})/h).$$

Now one has an expression for the variation of the function, the spatial derivative can be evaluated

$$\nabla p^{\mathrm{num}}(\mathbf{x}) = \sum_j \frac{m_j}{\rho_j} p_j \frac{\mathbf{x} - \mathbf{x}_j}{h|\mathbf{x} - \mathbf{x}_j|} \frac{\partial w}{\partial r}((\mathbf{x}_j - \mathbf{x})/h).$$

This approximation to the derivative suffers large $O(N^{-1/2})$ statistical fluctuations multiplying the mean value $p(\mathbf{x})$. These fluctuations can be eliminated by subtracting $p(\mathbf{x})$ from $p_j$ and by renormalising. Thus we have for the pressure gradient at fluid particle $i$,

$$\left(\nabla p^{\mathrm{num2}}\right)_i = \frac{1}{C} \sum_j \frac{m_j}{\rho_j} (p_j - p_i) \frac{\mathbf{x}_i - \mathbf{x}_j}{h|\mathbf{x}_i - \mathbf{x}_j|} \frac{\partial w}{\partial r}((\mathbf{x}_j - \mathbf{x}_i)/h)$$

$$\text{where} \quad 3C = \sum_j \frac{m_j}{\rho_j} (\mathbf{x}_j - \mathbf{x}_i) \cdot \frac{\mathbf{x}_i - \mathbf{x}_j}{h|\mathbf{x}_i - \mathbf{x}_j|} \frac{\partial w}{\partial r}((\mathbf{x}_j - \mathbf{x}_i)/h).$$

There is a similar expression for the divergence of the velocity field at a fluid particle.

Smoothing the pressure and the velocity fields over a distance $h$ with a symmetric smoothing function introduces an error $O(h^2)$ in the bulk. Within a distance $h$ from a boundary, the smoothing operation is no longer symmetric, which increases the errors to $O(h)$. Note that the renormalised expressions, $p^{\mathrm{num2}}$ and $\nabla p^{\mathrm{num2}}$, automatically compensate for the truncated volume near the boundary.

While the method is meshless and there is no need to structure the information stored, for efficient computation of the above sums it is useful to keep and update a list for each particle of nearby particles, so reducing an $O(N^2)$ process to an $O(N)$ process.

Boundary conditions are not simple to impose on the meshless method. For a fixed impenetrable boundary, extra image points can be introduced within a distance $h$ outside the boundary. These image points are given a velocity with the normal component reversed, to make that component vanish on the boundary. The pressure of the image point is given the same value, to make the normal component of the pressure gradient vanish on the boundary and hence the normal component of the acceleration vanish. For a prescribed impenetrable boundary moving at a constant velocity $V$ into the fluid, the normal component of the velocity of the image point needs to be set to $2V - v_i$, and the pressure set to the same value. If the impenetrable boundary is accelerating at $A$ into the fluid, then the pressure at the image point needs to be $p_i - 4\rho_i A \delta n_i$, where $\delta n_i$ is the distance of the point from the boundary.

Freely moving surfaces are more difficult, because it is part of the problem to determine the velocity and acceleration of the surface. There is also a question of deciding where the surface of the fluid is. At a crude level with $O(h)$ errors, one can just use the renormalised expressions with the points within the fluid and with no image points. This treatment of ignoring the lack of any information beyond the boundary does not properly impose the correct pressure on the boundary. To impose a pressure on the boundary one does need image points.

The dynamics is that of a compressible gas with sound waves propagating at speed $c = \sqrt{\gamma p/\rho}$. Resolving these waves imposes a CFL condition on the size of the time-step $\Delta t < h/c$. The dynamics is additionally inviscid, so that the sound waves propagate without dissipation. Small numerical errors are therefore conserved and can quickly swamp the real solution, i.e. there are the normal numerical problems of hyperbolic systems. Shockwaves may also occur.

There are extensions of the method described above to include viscosity. Sometimes just that part of viscosity to resolve the shockwaves is included. Viscosity requires second-order spatial derivatives to be evaluated, which are more difficult than the first derivatives.

In summary, the smooth particle method has the advantage of being meshless and also treats inertia well. In my opinion, these advantages are totally outweighed by the problems of achieving any accuracy at a low cost in storage and CPU time due to the need to find the pressure gradient and divergence in the velocity.

# Further reading

*Discrete element method simulations for complex granular flows* by Y. Guo and J. S. Curtis in Annu. Rev. Fluid Mech. (2015) **47**, 21–46.

*Immersed boundary methods* by R. Mittal and G. Iaccarino in Annu. Rev. Fluid Mech. (2005) **37**, 239–261.

*Lattice Boltzmann method for fluid flows* by S. Chen and G. D. Doolen in Annu. Rev. Fluid Mech. (1998) **30**, 329–364.

*Lattice-gas models of phase separation: interfaces, phase transitions, and multiphase flow* by Daniel H. Rothman and Stéphane Zaleski in Rev. Mod. Phys. (1994) **66**, 1417.

*Numerical simulations of particulate suspensions via discrete Boltzmann equation. Part I. Theoretical foundation* by A. J. C. Ladd in J. Fluid Mech. (1994) **271**, 285–309.

*Simulation method for particulate flows and concentrated suspension* by M. Maxey in Annu. Rev. Fluid Mech. (2017) **49**, 171–193.

*Smooth Particle Hydrodynamics and its diverse applications* by J. J. Monoghan in Annu. Rev. Fluid Mech. (2012) **44**, 323–346.

*Stokesian dynamics* by J. F. Brady and G. Bossis in Annu. Rev. Fluid Mech. (1988) **20**, 111–151.

# 17

## Wavelets

Wavelets have been particularly successful in compressing the data in an audio signal and in a visual image. In fluid mechanics, they have been used to represent numerical and experimental observations of turbulence and so reveal some of the structure in the turbulence. They have been suggested as an economical approach to computing turbulence, but so far without impact.

Finite differences give a localised representation of a function, whereas spectral methods give a global representation. Each has its merits and drawbacks. Finite differences are good at representing discontinuities. They are poor however at representing oscillations, where a single cycle needs at least eight points to give the most basic shape. On the other hand spectral methods represent oscillations well, but discontinuities poorly. For example, a discontinuity in the function $f(x)$ would give spectral amplitudes varying as $\tilde{f}(k) \propto 1/k$. While here the amplitude $\tilde{f}(k)$ is nonzero at each wavenumber $k$, nowhere in $f(x)$ could one find a wave with wavelength $2\pi/k$. (This is a warning not to be misled by the inertial $k^{-5/3}$ spectrum of homogeneous isotropic turbulence to expect to find a sequence of eddies within eddies.)

Wavelets sit between a local representation by finite differences and a global representation of spectral methods. Wavelets combine the advantages of both while mostly avoiding their problems. Wavelets are localised waves.

Think of a musical tune: a sequence of notes of different amplitudes, different frequencies and different durations. A Fourier analysis would not see the finite duration of each note. Finite differences would require eight points per cycle, lots of data for a musical note perhaps lasting tens of cycles. A musical score is so much more economical. To some extent wavelets mimic the musical score.

There are many possible functions that have been used for wavelets.

217

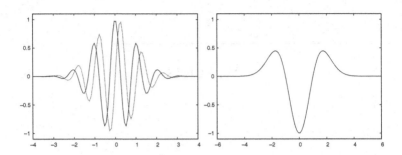

Figure 17.1 (a) The Morlet wavelet with $k = 6$. (b) The Marr Mexican hat wavelet.

Figure 17.1a shows the real and imaginary parts of the Morlet wavelet

$$\psi(x) = e^{ikx}e^{-x^2/2},$$

with wave-number $k = 6$. This wavelet was introduced by J. Morlet in the 1980s to analyse seismic data. Figure 17.1b shows the Mexican hat wavelet of D. C. Marr,

$$\psi(x) = \frac{d^2}{dx^2}e^{-x^2/2} = (x^2 - 1)e^{-x^2/2}.$$

These wavelets decay rapidly at large $x$. Their Fourier transform also decays rapidly at large $k$. All wavelet functions should be fairly well localised in space and in Fourier space. Here 'localised' can mean vanishing outside a finite region, although it is not possible for this to happen both in real and Fourier space for one function.

Often it is useful for the wavelet to have zero mean value, and even several moments vanishing. The above Morlet wavelet can be modified to make the mean vanish

$$\psi(x) = (e^{ikx} - e^{-k^2/2})e^{-x^2/2}.$$

For $k = 6$, as in Figure 17.1a, this modification is hardly detectable.

## 17.1 Continuous Wavelet Transform

From the basic wavelet function, called the *mother wavelet*, a family of similar wavelets is generated by *translation* through $b$ and *dilation* through a stretch of the $x$-scale by a factor $a > 0$,

$$\psi_{a,b}(x) = a^{-1/2}\psi\left(\frac{x - b}{a}\right).$$

The prefactor of $a^{-1/2}$ ensures that the integrals of the square of the modulus of the wavelets are equal, normally set to equal 1.

For a function $f(x)$, its wavelet components are found by the Continuous Wavelet Transform

$$f_{a,b} = \int \psi_{a,b}^*(x)f(x)\,dx,$$

where the star denotes the complex conjugate. This transform gives the analysis of the function into wavelets. From these components, the original function can be reconstructed through the inverse transform

$$f(x) = \frac{1}{C_\psi} \int f_{a,b}\psi_{a,b}(x)\frac{dadb}{a^2}.$$

This inverse transform gives the synthesis of the function from the wavelets. The constant $C_\psi$ is given in terms of the Fourier transform $\widetilde{\psi}(k)$ of the wavelet function $\psi(x)$

$$C_\psi = \int \frac{|\widetilde{\psi}(k)|^2}{|k|}\,dk.$$

For this integral to converge, it is necessary for the mean value of $\psi(x)$ to vanish, so that $\widetilde{\psi}(k) = 0$ at $k = 0$.

The inversion formula can be established by noting that the wavelet transform is a convolution of two functions, whose Fourier transform is therefore the product of the Fourier transforms of the two functions,

$$\tilde{f}_a(k) = \int f_{a,b}e^{-ikb}\,db = a^{1/2}\widetilde{\psi}^*(ka)\tilde{f}(k).$$

Similarly there is a convolution in the inversion formula, so taking Fourier transforms

$$\int \int f_{a,b}\psi_{a,b}(x)\frac{dadb}{a^2}e^{-ikx}\,dx = \int \tilde{f}_a a^{1/2}\widetilde{\psi}(ka)\frac{da}{a^2}$$
$$= \int \widetilde{\psi}^*(ka)\widetilde{\psi}(ka)\frac{da}{a}\tilde{f}(k) = C_\psi \tilde{f}(k).$$

Useful for applications to differential equations, the derivative of the wavelet component with respect to the translation variable is equal to the wavelet component of the spatial derivative of the function

$$\frac{\partial}{\partial b}f_{a,b} = \left(\frac{\partial f}{\partial x}\right)_{a,b}.$$

## 17.2 Discrete Wavelet Transform

The Continuous Wavelet Transform is for the theoretical study of wavelets. For numerical applications, one typically has a function given at a finite number of data points. As with the Discrete Fourier Transform described in §6.7 in the chapter on spectral methods, one needs a discrete version of the wavelet transform.

Consider a function $f(x)$ known at $N$ equispaced data points on the unit interval $[0, 1]$, i.e. at $x_k = k/N$ with $k = 0, \ldots, N - 1$. Write $f_k = f(x_k)$. To avoid problems at the boundary, consider periodic functions and work on the circle. For the Discrete Wavelet Transform, the number of data points should be a power of 2, i.e. $N = 2^n$.

The continuous range of dilations and translations is restricted to a discrete set, to a sequence of frequency doubling and integer translations,

$$\psi_{i,j}(x) = 2^{i/2}\psi(2^i x - j) \quad \text{for} \quad i = 0, \ldots, n - 1 \quad \text{and} \quad j = 0, \ldots, 2^i - 1.$$

There is thus one wavelet $\psi_{0,0}(x) = \psi(x)$ which covers the entire interval $[0, 1]$. There are two wavelets $\psi_{1,0}(x) = \sqrt{2}\psi(2x)$ and $\psi_{1,1}(x) = \sqrt{2}\psi(2x - 1)$ which cover the two half intervals $[0, \frac{1}{2}]$ and $[\frac{1}{2}, 1]$ respectively. This continues down to the finest level with $2^{n-1}$ wavelets covering the $2^{n-1}$ finest subintervals. Thus there are a total of $1 + 2 + \cdots 2^{n-1} = N - 1$ wavelets for the $N$ data points. This structure of refinements gives a *'multiscale representation'*.

The wavelet components of a function $f(x)$ are calculated

$$f_{i,j} = \frac{1}{N} \sum_k \psi_{i,j}(x_k)f(x_k).$$

Near to a boundary, some of the wavelets will be nonzero at some points outside the unit interval $[0, 1]$. Values of the periodic extension of $f(x)$ need to be used at these points outside.

If the wavelet function $\psi$ is nonzero only in the unit interval, then calculating the wavelet component $f_{0,0}$ is a sum over all $N$ points, $f_{1,0}$ and $f_{1,1}$ are each sums over $N/2$ points, down to the finest $f_{n-1,0}, f_{n-1,1}, \ldots$ which involve only two points. Thus calculating all the wavelet components is an $O(N \log_2 N)$ operation. Typically the wavelet function $\psi(x)$ vanishes or is negligibly small outside a longer finite range. The number of operations to calculate the wavelet components increases proportional to the length of this range.

In principle, one could use most continuous wavelet functions for the discrete version. However, there are significant advantages in using the few special wavelets which are orthogonal in the discrete version

$$\frac{1}{N} \sum_k \psi_{i,j}(x_k)\psi_{l,m}(x_k) = \delta_{il}\delta_{jm}.$$

With this orthogonality, the inversion formula is simply

$$f(x_k) = \sum_{i,j} f_{i,j}\psi_{i,j}(x_k).$$

There is a technical complication here with the mean value. The wavelets have zero mean value, and so the above inversion formula will only work for functions with zero mean. How this technical issue can be handled is discussed in the next section.

There are several families of discretely orthogonal wavelet functions: the Haar wavelet, Sinc wavelet, Meyer wavelet, Battle–Lemarié wavelet, Daubechies wavelets, symlets and Coiflets.

Data compression and denoising are achieved by setting to zero the wavelet coefficients smaller than a cutoff.

## 17.3 Fast Wavelet Transform

Because the wavelets are local, calculation of the wavelet components takes $O(N \log_2 N)$ operations rather than a naive estimate $O(N^2)$. There is however a faster $O(N)$ Fast Wavelet Transform for some orthogonal discrete wavelets. Understanding how the Fast Wavelet Transform works takes one further into the theory of wavelets than is appropriate for the brief survey of this chapter. Hence this section just gives a hint, starting with a simple example.

The Haar wavelet is a square wave

$$\psi(x) = \begin{cases} 1 & \text{if } 0 \le x < \frac{1}{2}, \\ -1 & \text{if } \frac{1}{2} \le x < 1, \\ 0 & \text{otherwise.} \end{cases}$$

As the range is a single interval, it is clear that each wavelet $\psi_{i,j}(x)$ is orthogonal to all other wavelets in different dilations, and is also orthogonal to all translations in the same dilation.

Consider the simple case of $N = 8 = 2^3$ points. The wavelet components are easily calculated,

$$f_{0,0} = \tfrac{1}{8}(f_0 + f_1 + f_2 + f_3 - f_4 - f_5 - f_6 - f_7),$$

$$f_{1,0} = \tfrac{\sqrt{2}}{8}(f_0 + f_1 - f_2 - f_3), \quad f_{1,1} = \tfrac{\sqrt{2}}{8}(f_4 + f_5 - f_6 - f_7),$$

$$f_{2,0} = \tfrac{1}{4}(f_0 - f_1), \quad f_{2,1} = \tfrac{1}{4}(f_2 - f_3), \quad f_{2,2} = \tfrac{1}{4}(f_4 - f_5), \quad f_{2,3} = \tfrac{1}{4}(f_6 - f_7).$$

Two remarks can be made. First, these seven components cannot represent the eight values of the function. What is missing is the mean value, which can be accommodated by introducing the sum

$$f_{0,0}^{\phi} = \tfrac{1}{8}(f_0 + f_1 + f_2 + f_3 + f_4 + f_5 + f_6 + f_7).$$

With this extra, the inversion formula works, e.g.

$$f_0 = f_{0,0}^{\phi}\phi_{0,0}(0) + f_{0,0}\psi_{0,0}(0) + f_{1,0}\psi_{1,0}(0) + f_{2,0}\psi_{2,0}(0),$$

with   $\phi_{0,0}(0) = \psi_{0,0}(0) = 1, \quad \psi_{1,0}(0) = \sqrt{2}$   and   $\psi_{2,0}(0) = 2.$

The second remark is that the sum $(f_0 + f_1)$ occurs in $f_{1,0}$, $f_{0,0}$ and $f_{0,0}^{\phi}$. A economy is available by not repeatedly adding the same two values together. The same can be said of other pairs.

The above two issues of the mean value and avoiding repeating additions are solved by introducing the *scaling function* $\phi(x)$, which for the Haar wavelets is simply a constant

$$\phi(x) = \begin{cases} 1 & \text{if } 0 \le x < 1, \\ 0 & \text{otherwise.} \end{cases}$$

As with the wavelets, dilations and translations of the basic scaling function are used

$$\phi_{i,j}(x) = 2^{i/2}\phi(2^i x - j) \quad \text{for} \quad i = 0,\ldots,n-1 \quad \text{and} \quad j = 0,\ldots,2^i - 1.$$

Components of the function $f(x)$ are also taken with respect to this family of scaling functions,

$$f_{i,j}^{\phi} = \frac{1}{N} \sum_k \phi_{i,j}(x_k)f(x_k).$$

It can be seen immediately that the single component $f_{0,0}^{\phi}$ with respect to the scaling functions is all that is needed to fix the issue of the mean value of $f(x)$ which could not be represented by the wavelets alone.

The other issue of avoiding repeating unnecessary additions, and thereby finding a Fast Wavelet Transform, can now be explained. One starts at the finest level $i = 2^{n-1}$, in our simple example

$$f_{2,0} = \tfrac{1}{4}(f_0 - f_1) \quad \text{and} \quad f_{2,0}^{\phi} = \tfrac{1}{4}(f_0 + f_1),$$

and similarly for all the other $f_{2,j}$ and $f_{2,j}^\phi$. At the next level one uses these $f_{2,j}^\phi$,

$$f_{1,0} = \tfrac{1}{\sqrt{2}}(f_{2,0}^\phi - f_{2,1}^\phi) \quad \text{and} \quad f_{1,0}^\phi = \tfrac{1}{\sqrt{2}}(f_{2,0}^\phi + f_{2,1}^\phi),$$

and similarly for all the other $f_{1,j}$ and $f_{1,j}^\phi$. And then at the top level in this simple example

$$f_{0,0} = \tfrac{1}{\sqrt{2}}(f_{1,0}^\phi - f_{1,1}^\phi) \quad \text{and} \quad f_{0,0}^\phi = \tfrac{1}{\sqrt{2}}(f_{1,0}^\phi + f_{1,1}^\phi).$$

This organisation of the calculations has avoided all the unnecessary repeating of additions, and requires just $O(N)$ operations.

Keeping with the simple example of Haar wavelets, the *Inverse Fast Wavelet Transform* can be explained. One starts with wavelet components $f_{i,j}$ and the coarsest scaling function component $f_{0,0}^\phi$. From the wavelet component $f_{0,0}$ and scaling function component $f_{0,0}^\phi$, one can form the scaling function components at the next finest level, $f_{1,0}^\phi$ and $f_{1,1}^\phi$, by working backwards the formulae above. From the wavelet components $f_{i,j}$ and scaling function components $f_{i,j}^\phi$, one can calculate the scaling function components $f_{i+1,j}^\phi$, and so on to the finest level to recover the function values. This calculation requires $O(N)$ operations.

A curious feature of the Fast Wavelet Transform and its inverse transform is that neither use the details of the scaling and wavelet functions. Instead they evaluate the components of the next level from the values of the components at the current level, with just two so-called *filter coefficients* $\pm\frac{1}{\sqrt{2}}$. This provides the generalisation from Haar wavelets to other wavelets.

At any $l$th stage, the partial sum

$$\sum_{j=0}^{2^l-1} f_{l,j}^\phi \phi_{l,j}(x)$$

represents all the coarser scale variations of the function which have not been described by wavelets at the scale of $l$ and finer, as in

$$\sum_{i \geq l, j} f_{i,j} \psi_{i,j}(x).$$

In the language of signal processing, the Fast Wavelet Transform is seen as a sequence of frequency filters, with a high-pass filter extracting the wavelet component and a low-pass filter extracting the remaining scaling function component. The sequence of filters starts at the highest frequency. The remaining scaling function component is then fed into the next filter which separates at half the frequency, etc.

## 17.4 Daubechies wavelets

From the idea that all the scaling functions $\phi_{i,j}$ must be able to be expressed as a linear combination of the scaling functions at one finer scale,

$$\phi_{i,j}(x) = \sum_k h_k \phi_{i+1,j+k}(x),$$

one becomes interested in special cases where the sum contains a small number of terms. The Daubechies D-2 has just four, so that

$$\phi(x) = \sqrt{2}\,(h_o\phi(2x) + h_1\phi(2x-1) + h_2\phi(2x-2) + h_3\phi(2x-3))\,.$$

Constraints of orthogonality, normalisation and some vanishing moments require

$$h_o = \frac{1+\sqrt{3}}{4\sqrt{2}}, \quad h_1 = \frac{3+\sqrt{3}}{4\sqrt{2}}, \quad h_2 = \frac{3-\sqrt{3}}{4\sqrt{2}}, \quad h_3 = \frac{1-\sqrt{3}}{4\sqrt{2}}.$$

The function that satisfies this functional equation can be found by iterating the expression starting from the Haar scaling function. The result is plotted in Figure 17.2. It vanishes outside the interval $[0, 3]$. The function is distinctly irregular, and is not suitable for use in solving differential equations! Fortunately it is never used, and instead one just uses the filter coefficients $h_k$ and $g_k$ in the Fast Wavelet Transform

$$f_{i,j}^\phi = \sum_k h_k f_{i+1,2i+k}^\phi \quad \text{and} \quad f_{i,j} = \sum_k g_k f_{i+1,2i+k}^\phi,$$

where the coefficients $g_k$ are the $h_k$ backwards with alternating signs

$$g_0 = -h_3, \quad g_1 = h_2, \quad g_2 = -h_1 \quad \text{and} \quad g_3 = h_0.$$

The Daubechies D-2 wavelet satisfies

$$\psi(x) = \sqrt{2}\,(g_o\phi(2x) + g_1\phi(2x-1) + g_2\phi(2x-2) + g_3\phi(2x-3))\,.$$

This is also plotted in Figure 17.2.

There is a good Wavelet Toolbox in MATLAB.

## Further reading

*Wavelet methods in computational fluid dynamics* by K. Schneider and O. V. Vasilyev in Annu. Rev. Fluid Mech. (2010) **42**, 473–503.

*Wavelet transforms and their applications to turbulence* by M. Farge in Annu. Rev. Fluid Mech. (1992) **24**, 395–457.

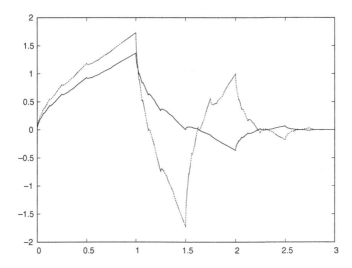

Figure 17.2 Daubechies-2 wavelet (curve with larger excursions) and scaling function.

# Index

Printed in the United States
By Bookmasters